Community Disaster Vulnerability

Community Disaster Vulnerability

Editor

Sidharth Saini

scitus
academics

Community Disaster Vulnerability

Edited by **Sidharth Saini**

Printed in 2017

ISBN: 978-1-68117-123-4

Library of Congress Control Number: 2015936545

© 2016 by

SCITUS Academics LLC,
616, Corporate Way, Suite 2, 4766,
Valley Cottage, NY 10989

www.scitusacademics.com

Contents

vi

Preface

Community disaster vulnerability is apparent that there is a need to integrate disaster mitigation and risk reduction into disaster recovery, economic and community development, and environmental policy and management. To reach these goals agencies and disciplines should work together, share knowledge and consider pre-planning strategies with the goal of increasing disaster resiliency and the overall economic health of their community. This book is intended to provide conceptual framework and empirical evidence of the factors contributing to disaster recovery and sustainable economic development. This is a unique feature of the book. Social vulnerability is one dimension of vulnerability to multiple stressors and shocks, including abuse, social exclusion and natural hazards. Social vulnerability refers to the inability of people, organizations, and societies to withstand adverse impacts from multiple stressors to which they are exposed. These impacts are due in part to characteristics inherent in social interactions, institutions, and systems of cultural values.

Editor

Socio-Environmental Vulnerability and Disaster Risk Reduction: The Role of Espírito Santo State (Brazil)

Raquel Otoni de Araújo[I] and Teresa C. da Silva Rosa[II]

[I]Mestrado em Sociologia Política (UVV, 2014). Universidade Vila
[II]Doutorado em Sócio-économie du développement (EHESS, 2005). Universidade Vila

ABSTRACT

Rainfall is a natural event having significant impact in the state of Espírito Santo (Brazil), resulting in floods and mass movements, damaging communities in a historical situation of socio-environmental vulnerability. The State Civil Defense is mobilized towards risk mitigation through the State Civil Defense Department (CEDEC) and through the development of the State Protection and Civil Defense Plan, coordinating multiple social actors. This paper aims to analyze the acts and role of the State in response to extreme events, focusing on how vulnerability, risk and disaster in the urban environment are tackled by government policy. Furthermore, it aims to discuss the contemporaneous actions after Hyogo 2005-2015, focusing on the concept of resilient cities, given the relevance of prevention work and actions by local actors, governmental or not, in the process of risk and disaster reduction and of mitigating vulnerabilities, demonstrating the application of the global discussion in local situations.

INTRODUCTION

The disaster of January 2011 in Rio de Janeiro State's Região Serrana (mountainous region) served as a warning to the country. Dubbed a mega-disaster (Brasil, 2012), the event is considered one of the biggest "climate" and mass movement (landslide) disasters ever to have occurred in the country. Seven municipalities were affected; most notably Petrópolis, Teresópolis and Nova Friburgo, and 947 people lost their lives. Disasters brought about by heavy rains leading to landslides and flooding are recurrent in Brazil's south-eastern region, primarily in the coastal Atlantic Forest belt.

In recent years, disasters of considerable environmental proportions and social impacts in Brazil have been numerous. In 2011 alone, 795 "natural" disasters were officially recorded in 2,370 municipalities[i], leading to the deaths of 1,094 people and affecting 12,535,401 others. Sixty-five per cent (65%) of these were of hydrological origin (Brasil, 2012). This situation is aggravated during what are considered to be normal periods of tropical rainfall, which have significantly intensified in a short space of time.

[i] 1,247 located in the Southern Region and 569 in the country's Southeast (Brasil, 2012).

In climatological terms, Brazil's south-eastern region falls under the influence of the Atlantic Convergence Zone (ACZ) and the ocean itself, which determine its rainfall systems primarily in the summer months between December and March. Located in this region, the State of Espírito Santo was afflicted in 2013 by two intensive rainfall events: In March, entire districts of the Greater Vitória Metropolitan Region

(RMGV) - where the State capital is situated - were stranded, most notably in the city of Vila Velha when, for forty-eight hours, the city practically came to a standstill; in December of last year - just before Christmas - rainfall affected the whole state, particularly the RMGV and areas in the north of the state.

In this latter case, people were prevented from circulating safely on highways or even city streets across the state for a week. Of the 78 municipalities in the state, 54 were adversely affected, and a state of emergency declared in 45. 44,577 people had to leave their homes temporarily: 6,471 were housed in shelters, with a further 38,106 going to stay with friends and relatives. Twenty-four people died as a result of the event. Highways were partially destroyed or flooded, the state capital›s airport was closed for several days and cities partly deluged. This situation mobilized thousands of capixabas (local name for the people of Espírito Santo State) over several days in campaigns for donation of food provisions and clothing for those affected[ii], along with four tons of medicines sent by the federal government.

[ii]G!. Veja a situação das rodovias estaduais e federais no ES. Available at <http://g1.globo.com/espirito-santo/noticia/2013/12/veja-situacao-das-rodovias-estaduais-e-federais-no-es.html>. Accessed on 18 november, 2013.

A common factor in recent disasters in Brazil has been the link between disordered growth of cities, giving rise to environmental degradation and social exclusion, and extreme events which, as referred to above, are very heavy tropical rains falling in a short period of time. Typical in regions of late development, the processes of economic growth and urbanization lead to occupation of Permanent Preservation Areas (PPAs)[iii], without any kind of urban planning, evidencing the lack of state intervention.

[iii]Constituted in 1965 to preserve biodiversity and according to the Forestry Code, Permanent Protection Areas are areas of: forests and other natural vegetation situated on the banks of lakes or rivers; hilltops; sandbanks and mangroves; hillsides; tableland or plateau edges with slopes.

Almost always, those worst-hit by extreme events are families with a background of social-environmental vulnerability living in such areas after having been marginalized by the real-estate market at a time of large-scale development projects in RMGV (Mattos & Da-Silva-Rosa,

2011). Such vulnerable populations are repeatedly obliged to leave their homes, some temporarily and others definitively as a result of floods occurring in low-lying areas (as is the case with Vila Velha, ES), or landslides in hillside areas. In some cases, these people may lose their belongings or even family members.

What can be observed (not only in the case of RMGV, but in the country as a whole) is that various different actors are improving their response to emergency situations arising from intensive meteorological events - on the one hand, traditional government actors who seek to organize themselves to assist populations afflicted by disasters, primarily those known to be housed in at-risk areas, and on the other hand civil society, which has been preparing itself to achieve readiness for an emergency, mobilizing an entire network of volunteers to assist those made homeless through distribution of food and clothing.

More recently, such actors have been operating within the new paradigm introduced by Marco de Hyogo (2005-2015) in which prevention is the crucial focus in disaster risk reduction (DRR) (UNISDR, 2005). Such change of direction has brought about considerable changes in the approach to disasters. If - before this global benchmark came into being - disasters were faced only when they occurred, then in the new approach they are complexified because they are now understood as being implicated by different situations outside the disaster itself. This new paradigm brings the disaster under the social microscope without leaving aside the natural perspective, which demands a wider outlook in its approach. It is acknowledgement of the complex character of a disaster which forms the basis of necessary dialogue between the various actors with an end to finding a solution – it is also worth making mention of dialogue within the scientific community in respect of knowledge production, because disasters in the Anthropocene era should be studied by the different areas of learning.

Traditionally, Brazil has always responded to disasters from an emergency response point of view, explained by the country's Civil Defense background. The origin of the Civil Defense organization in the country goes back to the Second World War and is related to the concept of national security (Ministério de Integração Nacional, 2013). In the 1960s, disasters occurring due to heavy rain in the region of Caraguatatuba (coastal São Paulo State) and in the city of Rio de

Janeiro stimulated implementation of Regional Civil Defense Agencies (Ministério de Integração Nacional, 2013). In the same decade, the Ministry of the Interior was created with the aim of aiding victims of calamities, and in 1992 came to be known as the National Integration Ministry where the National Civil Defense Department (CEDEC) is housed today. Ironically, it was the mega-disaster in Rio de Janeiro State's Região Serrana (mountainous region) that is now considered to have been the paradigm shifting point in public policy on this theme. In 2012, the National Civil Protection and Defense Policy was signed, in which the idea of prevention – until that time completely peripheral to the whole risk-management process – was integrated.

Some Brazilian municipalities have been successful in their risk management work, promoting actions that aim to mitigate the socioenvironmental vulnerability and to minimize risk as much as possible. Belo Horizonte, capital of Minas Gerais state, received, in 2013, the Sasakawa Award for Disaster Reduction in 2013, from UNISDR; Campinas, a municipality from the state of São Paulo, has been highlighted, internationally, as a pioneer in the UNISDR's program "Resilient Cities".

With this scenario in mind, this text seeks to analyze the actions and role of the State in dealing with extreme events, looking to focus on the manner in which vulnerability, risk and disaster in urban environments are approached during government actions. The State of Espírito Santo is a case in point due to the impacts of recent extreme events in communities made vulnerable by urban spread, reflecting problems common to countries and/or regions with a late development process. This paper also aims to discuss contemporaneous responses post-Hyogo 2005-2015, with focus on the resilient cities concept, permeated by the relevance of preventive measures and the performance of local actors, whether governmental or otherwise, in the process of disaster risk reduction and mitigation of vulnerabilities, thereby demonstrating that the global discussion on the theme has been applied to local situations.

To this end, an analysis was conducted of public policy discourse on civil defense and protection at federal and state levels, with the Hyogo Framework for Action 2005-2015 serving as a reference. Based on the bibliography and debates occurring, particularly in international forums, the resilient cities concept will be discussed as a possible post-Hyogo action.

This paper is divided into two parts – one dealing with the discussion on vulnerabilities and disasters in the Anthropocene era and the other presenting data on the analysis of two official documents on protection and civil defense in Espírito Santo State.

VULNERABILITIES AND DISASTERS IN THE ANTHROPOCENE ERA

Based on the assumption that human activities are intensifying and leaving an ecological footprint on the planet of such significance to have caused a break in biogeochemical cycles, disasters would be characterized by being truly complex because they would be breaking down the "limits"[iv] between the natural and the human which characterize the Anthropocene era (Bonneuil & Fressoz, 2013).

[iv]Constructed primarily during the Modern era, when men proclaimed their ability of dominating nature. Modern science has contributed, essentially through technology, to construction of spaces whose ecological, geomorphological or social characteristics were not respected.

This view, according to which the human being would be identified as the origin of events such as climate change, elevates the notion of disaster to a new level. In order to understand and deal with this within such a perspective, dialogue involving knowledge obtained from different areas of expertise and that which is produced on an interdisciplinary level become essential. Moreover, when viewed in this way, disasters are faced as the responsibility of humanity itself, revealing a given method of use (or abuse) of natural resources. As with other socio-environmental issues they have, in the natural-social system, their causes and effects.

As such, the notion of disaster is considered with a wider and more complex outlook, including it as a factor for criticism of current development standards and an intervening factor, in a medium to long-term scenario, of a model for ecological sustainability of development. It is therefore in the complex perspective, i.e. existing interconnectivities between the various dimensions which integrate the disaster situation, with the Anthropocene era as the backdrop, that disaster is dealt with in this paper.

The fact that disaster may be defined by different social actors renders the quest for its definition a more difficult task within the social perspective (Perry, 2005), as it brings out perceptions, meanings and interests beyond the historical context in which such actors are inserted. According to Perry (*id.*), the sociological perspective of disaster was initially brought to the fore by Fritz (*apud id.*) in the 1960s when he referred to the impact an event may have on a community. In Brazil, Valencio (2011) presents the idea of disaster as a tragic social event; in addition to material losses there are also, on the symbolic plane, emotional impacts and damage suffered by each individual. It is, however, the impacts on the social system - causing breakdown of social relationships and structures, which render the disaster an object of sociology - historically and spatially determined (Perry, 2005; Valencio, 2011).

Lindell (2011) breaks down the disaster into three phases: the pre-disaster, taking into consideration conditions prior to the event such as social vulnerability in a community; the trans-disaster, linked to the event itself; and the post-disaster, being the conditions for recovery. The author, in this way, draws attention to two important points in dealing with disasters: prevention with mitigation of vulnerabilities and enhancing a community's capacity for resilience, preparing it to face the event; and community recovery, including strengthening of resilience when a further event is imminent.

The discussion about vulnerability is present in many definitions of disaster. This relation started during the 1960s, a period called *hazard-disaster tradition*, which studies relate the environment, the natural event, the socioeconomic system, the people and the social relations with the disaster (Perry, 2007). The discussion about vulnerability was also highlighted during the 70s and 80s, mainly about the structural and contextual perspective of a disaster, focusing on the human and social relations (Gilbert, 1998). Therefore, vulnerability can be seen as multidimensional, including social, economical, environmental, psychological, and many other aspects, present in a community prior to a disaster.

Disaster is not, therefore, only about the moment at which it happens, but also the before-and-after when actions to stimulate and/or increase resilience are crucial to ensure that the community has the ability to react so as not to suffer such heavy impacts. It is understood that such actions are structured on the principle of sustainability. In the same way that DRR identifies it as its purpose, sustainability can be integrated as much into mitigation of vulnerabilities applied by the late-development model as into recovery and reconstruction.

The Brazilian Civil Protection and Defense organization take such view of disaster when the model of disaster treatment is broken down into prevention, preparation, response and reconstruction (Ministério da integração Nacional, undated.). In addition, the second notion has been disseminated in the country that each actor in the process should act in each of these phases, seeking teamwork and dialogue with peers and insertion into Contingency Plans, for example.

Based on the above, it is evident that the understanding of disaster assumed in this paper is subject to a social perspective, where responsibility falls upon political stances historically taken, assumed, or otherwise. Rooted as it is in the social system, a disaster is a social phenomenon (and not just physical as assumed for so long), because it is socially constructed (Quarantelli, 2005) on a physical base (considered here as environmental) involving aspects of geology, geomorphology and ecology of the area in which vulnerable communities are situated. An event, whether an earthquake or landslide occurring in an area uninhabited by humans, neither holds any interest for, nor will be the object of a sociological examination simply because it does not expose a breakdown of social relationships and/or institutions existing in human groups. Having said this, the environmental perspective cannot be overlooked, particularly in urban environments occupied in a disordered fashion in late-developing regions such as in the case of Espírito Santo State.

No disaster, therefore, is outside the community (Ribeiro, 1995) insofar as it reflects the presence or absence of historical decisions. On the other hand, such decisions enable a disaster to be mitigated if social actors adopt new political stances, which should occur based on a governance process in which all stakeholders have a voice. The disaster is therefore considered to be a phenomenon socially and historically constructed on an environmental base, to be known from that point on as a socio-environmental disaster.

It is during such decision-making (or its absence) that discussion is had on prevention and reconstruction of a dynamic capable of contributing to ecological and social sustainability of development, in which the idea of resilient cities[v] can take its place. This idea identifies sustainability as a development strategy capable of contributing to DRR based on ten essential steps[vi] for making cities resilient[vii]. This is a decision to be discussed and assumed by the community as a whole based on a governance process involving all actors and, primarily, the community. In this perspective, it is assumed that a disaster may present an opportunity for construction of a new project in society acting both in prevention and reconstruction. To this end, the disaster is seen as having the potential to produce policies (Guggenheim, 2014).

[v]UNISDR. Resilient Cities. Available at <http://www.unisdr.org/campaign/resilientcities/>. Accessed on 18 November, 2013.

[vi]UNISDR. Essentials.. Available at <http://www.unisdr.org/campaign/resilientcities/toolkit/essentials>. Accessed on 18 November, 2013/

[vii]Resilience is "The ability of a system, community or society exposed to hazards to resist, absorb, accommodate to and recover from the effects of a hazard in a timely and efficient manner..." (UNISDR, 2009)

THE ROLE OF THE STATE IN THE CASE OF ESPÍRITO SANTO

It is in this context that the role of the State emerges as one of the social actors to contribute to prevention and DRR. It is understood that, chiefly in a governance environment such as we see in modern times, all actors have an equally essential role to perform: local, regional and federal governments; community associations; NGOs and INGOs[viii]; the private sector; educational and research institutions, etc. Taking into consideration the current Anthropocene scenario, the principle of responsibility complexifies the role of the State in the process of providing social conditions which contribute towards the occurrence of a disaster. Social vulnerability reflects the lack of basic infrastructure and urban planning, non-observance of legislation on land occupation and use and lack of access to quality education and healthcare, among other aspects.

[viii]International Non-Government Organizations.

All such conditions are produced by the development model in which social inequality and exclusion of populations go hand-in-hand (Da-Silva-Rosa & Mattos, 2012), placing emphasis on the economic dimension and leaving other dimensions of the real situation aside. In this sense, the government political sphere has its portion of responsibility as, in the case of Brazil and Espírito Santo, the State was absent, for example, at the time of occupation of environmentally fragile areas such as hillsides, riverbanks or mangroves.

The disaster is not therefore a "neutral" event (Guggenheim, 2014), but reflects the absence of the State (or its consent) at a given moment when it did not act (and does not act, in some cases) to modify occupation of urban areas through public housing policies and urban and spatial planning. In this context, the aim of disaster sociology is to contribute to revealing political factors and social actors involved in the occurrence of disasters. In other words, seeking answers as to how the political sphere, through public policies or actions or omissions, contributes to creation of socio-environmental vulnerabilities, bearing in mind that it is these that give rise to disasters.

A situation of environmental injustice is produced within this context when vulnerabilities place at risk communities which, historically, live in circumstances of deprivation as is the case for populations excluded from the development process. Such approach to disasters, with the bias of socio-environmental vulnerabilities, highlights a further aspect for consideration: the fact that the disaster has an ethical dimension related to human rights (Sachs, 2008), as populations historically rendered vulnerable are those that – also according to the literature – are most probably under threat of suffering from the impacts of extreme events in addition to having a lower capacity for resilience and reconstruction. They end up as the *hostages* of a political body which, instead of prioritizing actions to make up for absence of or restricted access to basic services, refrained (and still refrains) from assuming its responsibility and respect for the principle of equality. Such principles, if included on the agenda, would strengthen and protect different communities with equality in such a manner that all could be ready to act from prevention to reconstruction in a context of socio-environmental disaster.

Such scenario of socio-environmental vulnerability construction is evident in the case of Espírito Santo (Map 1). In the 1970s, large-scale development projects attracted immigrants, primarily those expelled from declining coffee plantations in the state. At that particular moment, the Greater Vitória Metropolitan Region (RMGV) began to undergo increasingly accelerated urban expansion. Thereafter such process, coupled with the lack of urban planning and basic public services, exposed serious situations of poverty and socio-spatial segregation (Siqueira, 2010a & 2010b).

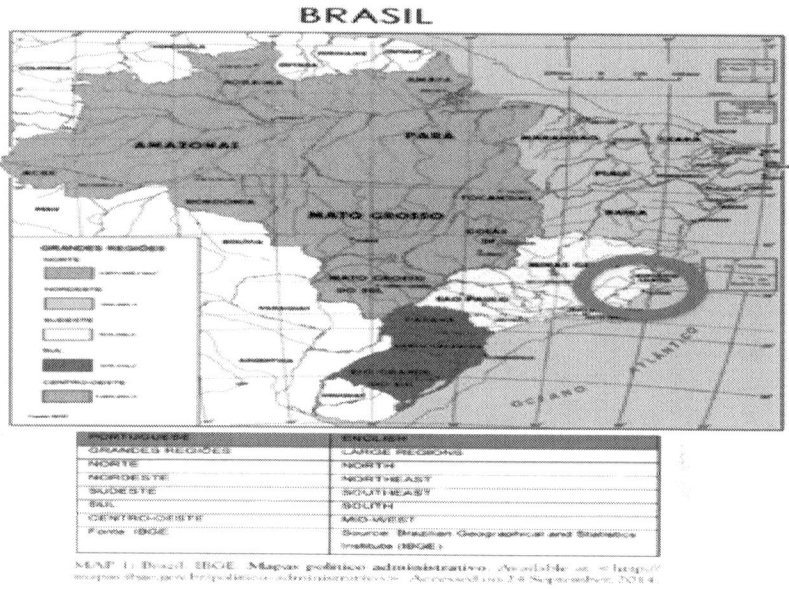

MAP 1: Brazil. IBGE. Mapas político administrativo.

The urban development process has, more recently, been stimulated by new development projects, primarily in the oil and gas sector and port construction, once again based on qualified labor drawn from outside the state and country due to the local workforce not being sufficiently qualified to operate in a technologically advanced environment.

Such process is currently intensified by new projects, under the auspices of exploitation of new natural resources, superimposing fresh problems on older ones and consequently aggravating socio-environmental vulnerability situations. If previously the lack of basic sanitation, health centers and good-quality schools could be "bypassed" with the solidarity of some actors (mainly churches), today such problems are chronic and more difficult to deal with in a city which is now consolidated, demanding another type of action from the State from a disaster-risk reduction point of view: prevention, preparation and reconstruction as a method of achieving sustainable development, with construction of resilient cities as the goal.

In 2013-2014, there were registered 74 occurrences, which resulted in either emergency situations or public calamity state. In those emergency situations, there were runoffs, droughts, mass movements, erosion, flooding and intense rains, being that runoffs were the most frequent event. The droughts were the only kind of event that resulted in a public calamity state, the were six registrations of them between 2013 and 2014 (DEFESA CIVIL DO ESPÍRITO SANTO, 2014). What has been observed in the case of recent disasters in Espírito Santo, as these listed, is the difficulty in taking public action in terms of prevention, preparation and sustainable reconstruction despite public policies – at both federal and state levels – paying more attention to such DRR management phases. This reminds us of what Guggenheim (2014, p. 11) talks about when he states that research on political spheres producing disasters highlights the inability of the State to take adequate risk reduction actions. In spite of the efforts being made across different agencies in Brazil, and in particular Espírito Santo, there is still much to do in respect of preparation, prevention and reconstruction concerning disasters.

There have been decades of a welfare-assistance mindset geared only for the emergency situation, when the risk-mitigation solution was merely technological and engineering-based, without taking into consideration the experience and capacity of the population falling victim to the disaster. The concept of disaster itself is rooted in Engineering, initially applied to physical and structural aspects; it then migrated to the social sciences, creating the requirement for expansion of the discussion beyond technical disciplines. It should be borne in mind that the Brazilian Civil Defense organization has its base in the state fire brigades, whose organizational structure is military, and that the very idea of the term is related to notions of militarization of national security – protection against an outside element (Gilbert, 1998); in other words far-removed from what, these days, could be called emergency management.

Nevertheless, Brazil has made advances in DRR, including the attempt to meet the commitment assumed on an international level with the Hyogo Framework for Action 2005-2015, established as part of the United Nations International Strategy for Disaster Reduction (UNISDR, 2005[ix]). Brazil is an active participant in international and regional discussions, in addition to promoting dialogue at national, state and municipal levels, integrating the various actors involved.

[ix]United Nations International Strategy for Disaster Reduction (UNISDR). 2005. Hyogo Framework for Action 2005-2015: building resilience of nations and communities to disasters.

PROTECTION AND CIVIL DEFENSE IN ESPÍRITO SANTO: UNDERSTANDING OF FRAGMENTED RISK

Two state-level references for DRR are discussed below: (1) the State Civil Defense and Protection Plan (PEPDEC), considered to be the contingency plan for ES, and: (2) Complementary State Law 694/2013 (CSL 694), which reorganizes the state Civil Defense and Protection System.

PEPDEC determines that the effects of a disaster be minimized and that "social normality" (sic) be re-established through actions for prevention, preparation and response. From a generalized point of view it meets the determination of the HFA; it is, however, possible to perceive some dissonance with the spirit of that framework.

PEPDEC actions focus more on response[x], revealing that it is much more of a contingency plan than a DRR management plan. The document addresses prevention and preparation, mainly in its description of the specific responsibilities of each agency and in presentation of risk mapping. It is also important to note that the guidelines presented in PEPDEC do not consider the issues of poverty reduction and community participation to satisfy local needs, as called for by HFA. On the other hand, risk is not understood in a complex way by administrators according to Araújo et al (2014). Some affirm that risk is only related to health issues, denying its complexity and reinforcing the idea of risk as a consequence of disaster and not as a pre-existing condition built over decades, as found in the HFA risk perspective. Thus, risk is not perceived as a situation prior to disaster itself, related to the historically-constructed social and environmental vulnerabilities of communities. This fragmented perception is present in both documents.

×See item 3.3 "Planning Assumptions".

It appears evident that the PEPDEC does not contemplate community contribution in elaboration or implementation of DRR actions. Local communities are merely treated as the population to be served by the actions, i.e. the community is not perceived as being one of the responsible actors and administrators in the process, as the HFA calls for. It should be recognized that at this level, two actors are not mentioned: the universities and the Instituto Jones dos Santos Neves, a state research agency.

In the two state laws above, risk is developed in a very narrow perspective in terms of its historical process, incorporating social vulnerability, ecological, geological and geomorphological aspects. For instance, no discussion is had about the development framework – an unequal and unsustainable development model focused on economic dimensions. Thus, causes are not really considered or treated in a complex manner focusing on natural and human elements. This means that the concept of risk in these documents is still very fragmented.

In PEPDEC 2012, risk appears much more as a factor to be mapped. What is observed in practice is disorganized mapping of risks – and vulnerabilities – among and between public agencies and other actors such as research institutions and universities. This will result in an overlap of actions (in this case mapping) with neither official mapping taking place, nor a single, coordinated source of data and information, even if drawn from different institutions and actors. Such overlap of actions confuses more than it contributes to implementation of DRR.

PEPDEC 2012 lists the duties of State departments in prevention, preparation and response, in particularly, incumbent on the State Civil Defense Coordination/CEDEC, among whose duties include support of the State Policy on Climate Change. When one analyzes actions to be put into practice by the different State departments, what is noticeable, primarily in respect of the sustainability principle, is a complete lack of terms which could refer to this category. For example, prevention actions, which could be aligned to the principle, prioritize response in a situation of emergency. As an example, one of the two actions of the Department for Sanitation, Housing and urban development provided for in PEPDEC (2012), seeks:

"to act in a preventive manner, with the support of municipal risk reduction plans, macro-drainage plans and execution of works for prevention and recuperation of damage caused by heavy rains, or recovery of water resources for prevention of drought" (p. 43).

Such vision reveals the still strong tendency towards a view of risk from a perspective of disaster and response rather than prevention, leaving aside the opportunity to implement sustainable actions. This is reinforced by CSL 694, demonstrating that the PEPDEC approach to risk is reproduced therein.

Another aspect to highlight is the difficulty in dealing with the notion of a consolidated city - as mentioned previously - raising the question of land-use and social inequality. In general, Brazilian laws on DRR are seen as being too strict because they do not consider this point. The same occurs with state-level policies on DRR. The consolidated city idea also reveals that laws are often established after irregular urban land occupation has occurred - in Brazil this means occupation of steep slopes, mangroves and riverbanks - areas that federal environmental law determines should be reserved for permanent preservation and in which it is illegal to build.

Bearing these points in mind, at least two questions arise from this analysis. Considering that HFA draws attention to the need to integrate risk into Poverty Reduction Strategies - related to development projects - how can DRR be integrated into development policies or planning when risk is not understood in its complexity? The second question is how to engage community participation in the sense of being aware of its needs if the population is not mentioned in such a document? Moreover, it must be borne in mind that the community to which the document would refer lacks basic education – hampering access to information required for DRR.

Despite the effort invested in establishing them, the two DRR reference documents for Espírito Santo State can be considered as inadequate due to their fragmented view of risk. This aspect may compromise integration of the notion of risk – while a complex category – in other policies, consequently missing the opportunity to fulfill the commitments assumed by the country in the international scenario. Such is the case of the Millennium Objectives, whose expiry date coincides with that of HFA in 2015; a year for review of what has been achieved in terms of the two international milestones, and for planning of what will be done going forward from then.

What is observed is an environment more prone to a lack of articulation between sector Public Policies, thereby compromising coordinated action prioritizing RRD, construction of sustainability and of resilient cities as mentioned above. Characterized in such a way, this present social environment appears more likely to exacerbate the situation of environmental injustice in which at-risk populations find themselves, doing nothing to contribute to the ethical dimension of respect for human rights. In other words, the current social-economical context does not promote, in any way, actions that develop and endorse human rights, especially towards those communities in a vulnerable situation.

In this sense, a need is apparent at state and municipal levels for improvements in discussion and awareness about risk, primarily among managers because they are very new to this area. This points to a lack of information on risk for those responsible for implementing actions or providing help to vulnerable communities. It also indicates a need for greater discussion about risk management involving public administrators and communities alike.

Finally, it is important to point out that communicability between the national, state and municipal levels of government is legislated for in Complementary State Law 694/2013, in terms of: (1) coordination and promotion regarding implementation of joint actions between state and municipal levels; (2) provision of information and support to the National Department for Protection and Civil Defense and concerning the occurrence of disasters and other civil defense activities. According to this law, the State government should promote development of public policies that help to create instruments for the joint execution of actions by the state and municipal Civil Defense and Protection Agencies which are able to enter into technical cooperation and financial agreements for the purpose of training.

FINAL CONSIDERATIONS

The DRR is one more strategy for implementation of an ecologically sustainable society project at a moment in which extreme climate-related events have occurred with greater frequency, exposing risk situations in which populations rendered vulnerable by a historical process live. To this end, DRR can be understood as an opportunity to enact sustainability principles, thus responding to the Millennium Objectives and climate change/sustainable development agendas.

The international benchmark regulating and guiding application of DRR actions is the Hyogo Framework for Action, in whose formulation Brazil played a part and has been noteworthy in its implementation nationwide. Municipalities such as Belo Horizonte (Minas Gerais) and Campinas (São Paulo) have been a reference. However some states, such as Espírito Santo, still need to update and/or amend their legislation and public policies to fall in line with the international discussion, integrating – in documented form – community and academic participation into actions of prevention, preparation, response and reconstruction. Although in their beginning, such practices are being applied, as observed during the 1st Inter-municipal Conference on Civil Defense and Protection (March 2014), and the 2nd State Conference on Civil Defense and Protection (April 2014), at which various actors discussed objectives and principles of the system and the different phases of action. These conferences happened as local and regional preparatory stages leading to the 2nd National Conference on Civil Defense and Protection (November 2014), which will discuss the new paradigms to the National System.

There is still apparent resistance to widening the theme to actors not in the public sphere, but there is a need for other state departments and agencies not directly involved in Civil Protection and Defense to be engaged in the discussion and practice of prevention, preparation and reconstruction, and not just with predefined actions to be executed in response. In this way the risk-management culture would be propagated in an integrated, horizontal manner whereby mitigation of vulnerabilities and risk reduction become the responsibility of all, with strengthening of communities and construction of resilient cities. Only in this way will Protection and Civil Defense actions, based on a risk-management strategy, become inter-departmental and interdisciplinary, developed by all corresponding departments and agencies with the involvement of other actors. Such action is increasingly necessary to ensure that DRR management does not lose its key focus - change to current standards, practices and development processes (UNISDR, 2013[xi]). To this end it is necessary to integrate DRR management into departmental public policies as a cross-fertilized theme.

[xi]This is the basic document upon which the future framework will be built through the formal preparatory process of the 3rd World Conference for Disaster Risk Reduction (Sendai, Japan, 2015).

Finally an essential point is noted: the requirement that development of discussions and actions, although perceived to be in the early stages in Espírito Santo State, not be altered in line with changes in public administration. Such practice has been a reality not just in this state but across Brazil: lack of continuity of actions. What is hoped is that the risk-management culture, mitigating vulnerabilities and building resilient communities do not become the agenda of one specific government or manager, but rather that it be something long-term, regardless of political administration changes – this is achievable through training and education of inter-departmental teams working, whether directly or indirectly, in socio-environmental disaster situations.

REFERENCES

1. ARAUJO, R. O. et all. Communicability between the national, state and municipal governments in the integration of the principles of the Hyogo Framework for Actions to reduce risks and disasters. Paper submitted to the Global Assessment Report on Disaster Risk Reduction, 2014.

2. BONNEUIL, C.; FRESSOZ, J. L'événement Antropocène. Paris, France: Le Seuil, 2013.

3. BRASIL. Ministério da Integração Nacional. Secretaria Nacional de Defesa Civil. Centro Nacional de Gerenciamento de Riscos e Desastres. Anuário brasileiro de desastres naturais: 2011. Brasília, Brasil, 2012.

4. DA-SILVA-ROSA, T.; MATTOS, R. Exclusion, Vulnerabilities and Climate Change. LASA - Latin American Studies Association. XXX

International Congress of the Latin American Studies Association. California, United States, 2012.

5. DEFESA CIVIL DO ESPÍRITO SANTO. SE/ECP Homologados. Available at <http://www.defesacivil.es.gov.br/(X(1)F(tig2NHj95UJJ07I5EVq0NUlNOoL-btTOYdDSsOXE03rEsi29Dg39K1VCpC951km6UqMeomfQ3-PGK0Q2i6ZnFt7mrYUGZOugaIcPtKGXW6PiLEB4kdtaPIZkV6920bdp2o5g-B3sVGf96dc-IfLRiW-K7U9Y1))/conteudo/SEECPHomologados/default.aspx>. Accessed on 24 September, 2014.

6. GILBERT, C. Studying disaster: changes in the main conceptual tools. In: QUARANTELLI. E. L. What is a disaster? Perspectives on the questions. New York, United States: Routledge, 1998.

7. GUGGENHEIM, M. Introduction: disasters as politics – politics as disasters. The Sociological Review, 62, 2014.

8. IBGE. Mapas político administrativo. Available at <http://mapas.ibge.gov.br/politico-administrativo>. Accessed on 24 September, 2014.

9. LINDELL, M. K. Disaster Studies. Sociopedia.isa. International Sociology Association, 2011.

10. MATTOS, R.; DA-SILVA-ROSA, T. Reestruturação econômica e segregação sócioespacial: uma análise da Região da Grande Terra Vermelha. I Seminário Nacional do Programa de Pós-Graduação em Ciências Sociais UFES, Vitória-ES. Anais do I Seminário Nacional do Programa de Pós-Graduação em Ciências Sociais UFES. Vitória, Brasil, 2011.

11. MINISTÉRIO DA INTEGRAÇÃO NACIONAL, Secretaria Nacional de Defesa Civil. Manual de planejamento em Defesa Civil, vol. 1. Brasília, Brazil, undated.

12. MINISTÉRIO DE INTEGRAÇÃO NACIONAL. Histórico da Defesa Civil. Available at <http://www.mi.gov.br/historico-sedec>. Accessed on 18 november, 2013.

13. PERRY, R. W. Disasters, definitions and theory construction. In Perry, R. W.; Quarantelli, E. L. (orgs.). What is a disaster? New answers to old questions. United States: International Research Committee on Disasters, 2005.

14. PERRY, R. W. What is disaster? In RODRÍGUES, H.; QUARANTELLI, E. L.; DYNES, R. R. Handbook of Disaster Research. Nova Iorque: Springer, 2007.

15. QUARANTELLI, E. L. What is a disaster? New answers to old questions. United States: International Research Committee on Disasters, 2005.

16. RIBEIRO, M.J. A construção de um modelo de análise das vulnerabilidades sociais dos desastres: uma aplicação à colina do castelo de S. Jorge. Territorium, vol. 13, 1995.

17. SACHS, W. Climate change and human rights. Development, vol. 51, no. 3, 2008.

18. SIQUEIRA, M.P.S. Os grandes projetos industriais: desenvolvimento econômico e contradições urbanas. Desenvolvimento Brasileiro: alternativas e contradições. Vitória, Brasil, 2010a.

19. SIQUEIRA, M.P.S. Industrialização e empobrecimento urbano: o caso da Grande Vitoria, 1950-1980. Vitória, Brasil: Grafitusa, 2010b.

20. UNISDR. Hyogo Framework Action: building the resilience of nations and communities to disasters. Geneva, Switzerland, 2005.

21. UNISDR. UN Special Representative of the Secretary-General for Disaster Risk Reduction. Proposed Elements for Consideration in the Post-2015 Framework for Disaster Risk Reduction, Geneva, Switzerland, 2013.

22. VALENCIO, N. A sociologia dos desastres: perspectiva para uma sociedade de direitos. Seminário estadual de emergências e desastres: estratégias latino-americanas de enfrentamento à questão. Vitória, Brazil, 2011.

Adding Value to Critical Infrastructure Research and Disaster Risk Management: The Resilience Concept

Claudia Bach[1], Sara Bouchon[2], Alexander Fekete[3],
Jörn Birkmann[1], and Damien Serre[4]

[1]United Nation University Institute for Environment and Human Security (UNU-EHS) Platz der Vereinten Nationen 1, 53113,

[2]Risk Governance Solutions S.r.l., Via Fratelli d'Italia, 7, 21052 - Busto Arsizio (VA),

[3]Institute of Rescue Engineering and Civil Protection, Cologne University of Applied Sciences/Fachhochschule Koeln,

[4]RESCUESolutions SAS, Bagneux, France damien.

ABSTRACT

In recent years, resilience has become a key term in disaster risk management (DRM). Its potential has been mainly discussed with respect to social-ecological systems as well as communities. With respect to Critical Infrastructures (CIs) however, resilience and vulnerability are often used without clear definition and reference to the DRM context. This paper aims to conceptualize vulnerability and resilience for the CI context. Building on socio-ecological approaches, the paper will outline the added value that a more stringent conceptualization of resilience offers for DRM of CIs. After an introduction of CIs and their meaning in the context of DRM (Section 1), the distinct features of the resilience concept and its application in different disciplines are presented (Section 2). Some of the governance challenges associated with the implementation of resilience strategies are presented (Section 3) before conclusions are drawn (Section 4).

INTRODUCTION

The recognition of risk as a social construct became one basis for the development of a certain stream of risk assessment methodologies and DRM approaches (e.g. Blaikie et al., 1994; Alexander, 2000; Birkmann, 2013). In this context, different terminologies and concepts such as vulnerability (Birkmann, 2013), sensitivity (Füssel & Klein, 2006), resilience (Paton & Johnston, 2000; Klein et al., 2003; Adger et al., 2005; Cutter et al., 2008) or adaptation and adaptive capacity (Pelling, 2011; Smit & Wandel, 2006) have been developed from related disciplines. Discussions with respect to their delineation, overlap and applicability are ongoing (Cutter et al., 2008; Cardona, 2011; Birkmann, 2013).

Definition and translation of these theoretical concepts into indicators and criteria form an important part of disaster risk assessments and are a priority of the Hyogo Framework for Action (UNISDR, 2007). In this respect, different spheres of interest have been identified that encompass economic, environmental and social dimensions (Cardona & Barbat, 2000; Birkmann, 2013; Cardona, 2011). Defined as "...an asset or part thereof... which is essential for the maintenance of vital societal functions, health, safety, security, economic or social well-being of people" (EC, 2008: Article 2a), Critical Infrastructure (CI) can be identified as a cross-cutting topic for all three spheres.

In parallel to these discourses in the DRM community, CIs have gained political importance in the wake of terror attacks in 2001 (World Trade Center), 2004 (Madrid) and 2005 (London) (FMIG, 2009; Her

Majesty the Queen in Right of Canada, 2009; HM Government, 2010). These events both shifted the focus of DRM activities and reshaped the CI context, through increasing awareness of the complexity and interrelatedness of infrastructures as socio-technical systems (e.g. Rinaldi et al., 2001; IRGC, 2006; Kröger, 2008; Serre et al., 2013) and the increasing likelihood of cascading effects reaching beyond geographical and functional borders (Boin & McConnell, 2007; Hémond & Robert, 2012; Lhomme et al., 2013).

The strengthening of infrastructures has been identified as an important field for disaster risk reduction (e.g. UNISDR, 2007). However, CI and DRM terminologies and methodologies have not fully been integrated, resulting in inconsistent labeling, conceptualization and implementation of disaster risk-related CI activities and governance approaches. Accordingly, it is the aim of this paper to apply methodological discussions on DRM conceptualizations to CIs, in order to underline the advantages of the resilience concept for this specific context as well as to discuss potential governance challenges.

FROM CRITICAL INFRASTRUCTURE PROTECTION TO RESILIENCE

In the context of DRM, the resilience concept is variously viewed as supplementary to (Gallopín, 2006), overlapping with (Cutter et al., 2008) or the flip-side of (Folke et al., 2002) existing concepts such as vulnerability. In the following, we will analyze the specificities offered by the resilience concept as well as its application in the development of strategies.

The Resilience Concept

The resilience approach was initially used in the fields of psychology (e.g. Garmezy et al., 1984; Rutter, 1985) and ecology (e.g. Holling, 1973), amongst others. In ecology, when considering systemic interactions, the term 'resilience' addresses the ability of ecosystems to absorb fluctuations while persisting. This was a departure from the traditional view that had equated the optimum ecological state with stability—a departure deemed necessary in order to address the behavior of nonlinear systems (Holling, 1973). Socio-ecological resilience research focuses on the relevance of renewal, reorganization and development (Holling, 2001; Gunderson & Holling, 2002; Berkes et al., 2008), arguing that resilience increases the likelihood for desirable pathways under changing and sometimes even unpredictable conditions (Walker et al., 2004; Adger et al., 2005; Folke, 2006). Accordingly, non-linear developments (which might also be generated through infrastructure breakdowns) became part of the analysis (Folke, 2006). The Resilience Alliance defined the term as:

"The ability to absorb disturbances, to be changed and then to re-organise and still have the same identity (retain the same basic structure and ways of functioning). It includes the ability to learn from the disturbance. A resilient system is forgiving of external shocks. As resilience declines, the magnitude of a shock from which it cannot recover gets smaller and smaller. Resilience shifts attention from purely growth and efficiency to needed recovery and flexibility. Growth and efficiency alone can often lead ecological systems, businesses and societies into fragile rigidities, exposing them to turbulent transformation. Learning, recovery and flexibility open eyes to novelty and new worlds of opportunity."[1]

[1]http://www.resalliance.org/index.php/resilience

Counter to this, the equilibrium approach to resilience played an important role in many disciplines and has substantially shaped natural resource and environmental management. Traditional engineering resilience approaches often focus on maintaining efficiency and the constancy of a system close to a single steady state (see Holling, 1996 and Table 1). This aspect can also be found in more recent engineering literature stressing the control over the system in order to avoid failure. Vugrin et al., (2010), for example, define CI resilience as:

"Given the occurrence of a particular disruptive event (or set of events), the resilience of a system to that event (or events) is the ability to efficiently reduce both the magnitude and duration of the deviation from targeted system performance levels" (p.82).

According to them, CI resilience comprises two main measurable components: a—the system impact, defined as the difference between general and actual (after event) performance; and b—the recovery effort, encompassing the resources required to restore the functioning to a pre-defined desirable performance level. This engineering driven approach thus neglects the potential for flexibility and change of the system. It relates resilience to the capabilities of systems or networks, elements often expressed in terms such as robustness, redundancy or others (Tierney & Bruneau, 2007). In these approaches, resilience assessments were and still are in many contexts addressing the physical conditions of systems while neglecting different aspects and phases of disaster management (see e.g. Hartong et al., 2008, Svensson 2008, Bompard et al., 2009, Rich et al., 2009, Gheorghe and Vamanu 2005 and 2008, Petit et al., 2011, Kröger and Zio 2011, Li et al., 2012).

Table 1: Concepts of resilience in the socio-ecological context

Resilience Concept	Characteristics	Focus on	Context
Engineering resilience.	Maintaining efficiency and constancy.	Deviation from actual performance (often also understood as robustness), recovery effort.	Vicinity to stable equilibrium.
Ecological/ ecosystem resilience and social resilience.	Buffering capacity, withstanding shock, maintaining function.	Persistence, absorb disturbance.	Multiple equilibria, stability landscape.
Socioecological resilience.	Interplay, disturbance and reorganization, sustaining and developing.	Adaptive capacity, transformability, learning, innovation.	Integrated system feedback, cross-scale dynamic interactions.

As indicated in Table 1 the resilience concepts used in the socio-ecological context allow for the consideration of systemic feedbacks and cross-scale dynamic interactions as well as (institutional) learning, which can also be transferred to CIs.

Infrastructure Protection: the Added Value of the Resilience Concept

Applying the concept of resilience as defined in the socioecological approach to CIs can be of great value. Shifting the focus away from the maintenance and equilibrium of the infrastructure system towards the delivery of system services and its external relations permits a better consideration of external effects and changes as well as interaction with other systems, in this case society. Gallopín (2006) suggests that the interaction of external and internal processes needs to be considered since stress can be triggered by changes in the system environment, by internal alterations, or both. With respect to CIs, a variety of system challenges can be identified. They include: the increasing (inter-)dependencies of and between infrastructure systems (e.g. Rinaldi et al., 2001); technological changes and the integration of smaller into larger systems, thus increasing system complexity and allowing for far-reaching disturbances (Kröger, 2008: 1781); the privatization of infrastructures (Gheorghe et al., 2006: xiv; Kröger, 2008: 178,); the liberalization of markets, leading to an increasing number of actors (Gheorghe et al., 2006: xi ff); and changes in demand patterns (Kröger, 2008; IRGC, 2010). Additionally, changes in system set-up and governance, and global changes including the increasing use of renewable energies, urbanization processes and demographic changes, also shape CI resilience.

Besides these changes, the resilience concept also allows for the consideration of unexpected events such as the 2004 boxing day tsunami, Hurricane Katrina in 2005 or the 2011 Tohoku earthquake and following tsunami. In some cases, unexpected disruptions may occur due to miscalculations in design. The Fukushima earthquake and tsunami for example were both larger than had been anticipated in the design of Japanese power plants (Bunn & Heinonen, 2011: 1580) representing a mismatch of design structures and the spectrum of plausible hazards. Incorporating the resilience concept in this case would ideally have led to the integration of 'safe failure' into the design structures, and a higher degree of flexibility to account for a diversity of hazards. In other cases, unexpected disruptions may be the result of cascading effects caused by increasing interdependencies and complexities (Rinaldi et al., 2001; IRGC, 2006; Boin & McConnell,

2007; Kröger, 2008; Hémond & Robert 2012; Lhomme et al., 2013): in this example, the dependency of tsunami height on the magnitude of the earthquake, and the interdependencies of the cooling system with electricity production, water pollution, back-up systems, blocked traffic routes and fire-brigade services. A DRM strategy informed by resilience would specifically design and implement measures that take the interlinkages between different infrastructures into account, as opposed to a DRM strategy that devises a separate strategy or analysis for each system.

Nevertheless, assessment methodologies and measures to address CI disruptions still mainly build on the notion of stability and robustness defined in specific scenarios (e.g. Greenberg et al., 2007; EC, 2009; Reed et al., 2009) while neglecting systemic changes and unexpected events. Although many assessments do encompass multi-hazard approaches, they are still creating blind spots and potential new vulnerabilities by not integrating resilience aspects into their assessments (Perelman, 2006). In this regard, resilience strategies need to incorporate uncertainties (or the so-called 'soft paradigm': Perelman, 2006). In its simplest form, this would mean having a "Plan B" in the event of failure (Tierney & Bruneau, 2007), or a range of options to be taken. Another strategy is to design 'safe failure' or 'graceful degradation' into systems, so that they continue to operate in the event of failure in one or more components (Tyler & Moench, 2012). Another approach is to openly encourage flexibility, for instance, encouraging people to embrace uncertainty rather than insisting on 100% certainty or even a predictable future and security.

CI/Human Interaction as Part of the Resilience Concept

Many approaches in the disaster risk reduction area are still sector-specific, focussing on the vulnerability of a particular type of system/CI (e.g. Hartong et al., 2008; Svensson, 2008; Bompard et al., 2009; Rich et al., 2009; Gheorghe & Vamanu, 2005, 2008; Petit et al., 2011; Kröger & Zio, 2011; Li et al., 2012). Although the respective research is valuable in order to learn more about the individual system characteristics and potential disaster risk reduction measures, the implications for society are often unclear. However, the operation of CIs determines the functioning of many societies (e.g. FMIG, 2009;

DHS, 2009; Cabinet Office, 2010); therefore, a broader perspective is required, that addresses the societal effects. The resilience concept offers the possibility to include societal aspects by taking into account the ability to absorb external shocks[2]. This is specifically relevant as the social effects of an infrastructure breakdown are mainly determined by the level of dependence on an uninterrupted supply, or by the level of preparedness (Toubin et al., 2014). Paradoxically, high levels of supply security lead to complacency within the population and thus an unpreparedness towards potential failures (FMIG, 2009; Reichenbach et al., 2008).

[2]See the Resilience Alliance website: http://www.resalliance.org/ index.php/ resilience

Taking electricity supply failure as an example, not all households and facilities are equally affected. While a shortfall of electricity supply can have life-threatening effects in hospitals (where emergency power supply might be insufficient) (Aghababian, 1994: 773; Klein et al., 2005: 343) and geriatric homes with patients dependent on artificial respiration, a household of middle-aged adults might be relatively unaffected. Additionally, preparedness levels, e.g. with respect to the availability of back-up facilities, will differ and thus influence overall CI-human resilience.

Development of Critical Infrastructure Resilience Strategies

Although the protection of strategically important facilities has always been an important part of national defense strategies (Hellström, 2007; Lauwe & Riegel, 2008), the beginning of the 21st century saw a change in the nature of perceived threats, with natural hazards and terrorism now the focus of security debates (Lauwe & Riegel, 2008). The importance of CI protection escalated through the 2001, 2004 and 2005 terror attacks in New York, Washington, Madrid and London, as well as in response to a number of disasters such as Hurricane Katrina in 2005 or the UK flooding in 2007. Against this changing landscape, societies have become highly vulnerable towards a broad and diffuse spectrum of possible threats (Brunner & Suter, 2008; Her Majesty the Queen in Right of Canada, 2009: 4; HM Government, 2010).

The first generation of policies addressing CI disruptions in this changed threat context was a set of Critical Infrastructure Protection (CIP) strategies. The focus on protection was mainly considered from an all-hazards perspective:

"...the objectives of the EPCIP [European Programme for Critical Infrastructure Protection] will be to continue to identify critical infrastructure, analyse vulnerability and interdependence, and come forward with solutions to protect from, and prepare for, all hazards." (EC, 2004: 8)

The protection was seen as the way to reduce, sometimes to totally eliminate, the vulnerabilities of critical infrastructure systems, mainly seen as the physical assets or components of an infrastructure. In this context, the vulnerability of CIs was defined as a "weakness in the system of the critical infrastructure in itself, which might be exploited, unintentionally or intentionally" (Bouchon, 2006: 80).

Nevertheless, during the last decade, CI disruption issues became more integrated into DRM approaches (see Figure 1). This increase in awareness on the importance of CI resilience, triggered by a variety of events such as the 2003 Northeastern blackout in the US and Canada, Hurricane Katrina in 2005, the 2007 UK flooding as well as the overall awareness of the potential effects of climate change (City of Cape Town, 2006; German Federal Government, 2008), was accompanied by a shift from direct protection and prosecution to a more systemic view of infrastructures in certain countries. It was characterized by the insight that rather than focussing on the protection of certain facilities, the safeguarding of the provision of services should be the primary aim. In particular, the Nordic countries focused on critical societal functions (Norwegian CIP Commission, 2006) or functions vital to society (Government of Finland, 2006).

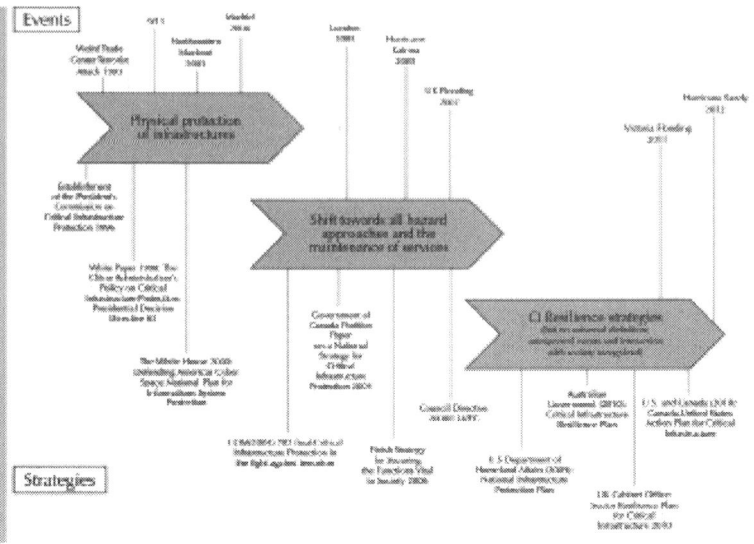

Figure 1: Development of Critical Infrastructure Resilience Strategies. Source: authors.

Some recent documents include resilience terminology in strategic approaches (City of New York, 2013; Cabinet Office, 2010; Scottish Government, 2011; Australian Government, 2010; DHS & Public Safety Canada, 2009; Her Majesty the Queen in Right of Canada, 2009; NIAC, 2009). However, in others the resilience approach is not referred to (e.g. EC, 2008) or is used in a misleading way in what should be referred to as protection strategies (e.g. The White House, 2013). This is astonishing not only since the scientific debate has been well developed, but because resilience-based approaches have been found to be substantially less expensive than investments into structural updates (de Bruijne & van Eeten, 2007: 24). Although resilience aspects such as education and training or specific response and recovery efforts are integrated into some strategies (e.g. DHS, 2009), societal resilience encompassing efforts and planning activities of communities and businesses is frequently neglected (Boin & McConnell, 2007: 54 ff; Pursiainen, 2007: 31 ff). Finally, most CI DRM strategies still fail to integrate insecurities and to explicitly address the potential unexpectedness of events. In this respect, the development of more comprehensive and integrated resilience approaches taking into account different spatial and content-related levels (e.g. changes in sociotechnical landscapes, patchworks of standards and regimes or action of individuals) are needed (Hellström, 2007). Although some guidelines take these aspects into account (e.g. TISP, 2006; TNO, 2011), a variety of challenges remain for regional, national and local strategies (Balsells et al., 2013; Toubin et al., 2014).

DISCUSSING CRITICAL INFRASTRUCTURE RESILIENCE IMPLEMENTATION CHALLENGES

Having looked into the development of CI resilience strategies, the variety of nomenclature is striking and increases the challenges in implementing resilience strategies such as 1—cooperation and communication among the multiplicity of stakeholders, 2—understanding system characteristics, and 3—the integration of citizens into resilience building.

Cooperation and Communication among the Multiplicity of Stakeholders

In the field of CI strategies, the main challenge raised by governance is related to the need to involve very different types of stakeholders: the population, public authorities (from different jurisdictions), private operators, operators from different sectors, the media, etc.. This need has been articulated in the concept of network governance proposed by Suter (2011). Each group of stakeholders has its own interests, and hence its own understanding of what it means to achieve resilience. Potential conflicts may arise, for instance when business interests are not compatible with public security. The role of the State with respect to CIs changes when liberalization or privatization occurs: it can no longer directly influence the setup and governance of the CI system, but rather has to focus on setting framework conditions for production processes and markets and subsequently organizing and moderating negotiations between different stakeholders (Abbate, 1999; de Bruijne & van Eeten, 2007; Monstadt, 2008; Toubin et al., 2014). The provision of CI services is, however, mainly in the hands of privately organized operators.

The collaboration between the different stakeholders requires adequate collaboration schemes, where each group of stakeholders feels that its own interests are taken into account (win-win situations). The need for cooperation was already identified by the President's Commission on Critical Infrastructure Protection (PCCIP) (1997). Such public-private partnerships (PPPs) are already operational, for instance, in Scotland, where the regional Critical Infrastructure Strategy is based on a Critical Infrastructure Partnership Framework between Government and those responsible for the critical assets "...to minimise disruption to any part of that infrastructure or to any of our communities living and working across Scotland" (Scottish Government, 2011: 27). In the case where the interests of each group of stakeholders are not taken on board, the strategy elaborated may compromise the achievement of the expected results, as in the example of the 2008 European Directive on European Critical Infrastructure (ECI). The Directive defined security measures to be implemented by the ECI operators; however, the operators had little input on the accuracy of these measures, whose implementation they had to pay for (Bouchon, 2011). This was one of the main factors triggering a revision of the Directive. Resilience measures should thus be the result of a participative process to ensure better acceptance.

In the case of Norway, CI resilience is achieved through measures implemented by responsible owners, taking into account the needs defined by their customers, and on the basis of goals, expectations and regulations defined by the responsible authorities, within a system of risk governance defined by the government (Thomassen, 2012). However, adequate governance models to achieve CI resilience in different contexts remain to be developed. The main challenges in this regard include: the identification of the right number of stakeholders (not too few, not too many); the types of collaboration process to be developed (e.g. protocols, informal discussions, exercises, etc.); and the degree of formalization of this process (policies with a normative goal vs. processes based on a voluntary, informal, or horizontal way of collaboration).

If CIR stakeholder collaboration is to prove efficient in addressing the resilience and protection issues (Bouchon & Dimauro, 2012;

Bouchon et al., 2012), it is fundamental that the interest of participants be maintained by taking into account their needs and perspectives, and also that the necessary funding be generated. New synergies and innovative cofinancing strategies, involving the participation of both public and private actors, need to be explored in this regard.

The adoption and the development of adequate technologies and communication systems are particularly relevant to CI Resilience collaboration processes: they are the key factors to secure the exchange of information. Large, monolithic and costly integrated platforms are not in line with the needs of authorities and operators. On the contrary, technological innovations should be directed to gain opportunities for economy of scale and investment sustainability, thanks to modular interoperable and reconfigurable solutions.

A key element of successful stakeholder collaborations is communication between the stakeholders (IRGC, 2006). Communication is principally a question of defining responsibilities, roles, duties and obligations with respect to specific procedures and crisis situations. The challenges of CI governancealready substantial in business as usual , scenarios against the background of liberalized markets and the privatization of the provision of services (Kröger, 2008)are only intensified in times of crisis when specific needs arise but responsibilities and cooperation potential remain unclear. For example, Hurricane Katrina showed that the disaster management mechanisms in place were not sufficient, despite the fact that new strategies had been implemented after 9/11 (Wise, 2006). Although the unexpected breakdown of communications technology played an important role in turning the natural hazard into a disaster situation, confusion about responsibilities and the interaction of departments and officials contributed significantly (ibid.: 304). Since communication always involves the exchange of information, it is important to address the type of information that can be exchanged (e.g. intelligence or commercially sensitive information) and technical questions related to the information exchange, such as protection measures, secure information sharing platforms and their limitations in DRM.

Finally, decision-making processes also pose a challenge to resilience building. This encompasses decision-taking as well as the financing, implementation and monitoring of decisions taken. It is also necessary to determine how decisions can be taken. For example, who should make decisions regarding resilience levels and acceptable risks? Ensuring the delivery of certain infrastructure services during times of crisis necessarily implies the prioritization of available resources; however, the criteria under which such prioritization processes can be organized, and who should make such decisions, remain unclear. Public discussion about possible limits of protection and target levels of risk can contribute to addressing this challenge and could serve as a stimulus for multi-stakeholder communication about risks (Fekete et al., 2012; Fekete, 2012).

Understanding System Characteristics

A second governance challenge that is closely related to communication between stakeholders is related to the collection of relevant information to characterize the CI systems and their interdependencies. Stakeholders need to understand their dependence on certain CI services as well as their own role in the functioning of other infrastructure services. Such system characteristics can be identified in a hazard independent manner with the help of scenarios.

Some cases show that building PPPs, as mentioned above, creates a trusted environment for information exchange that allows better understanding of the system characteristics. Examples have been developed in the Netherlands (Luiijf et al., 2003), Canada (Robert & Morabito, 2010), in Scotland, or in Lombardy, Italy (Bouchon et al., 2012). In this last example, transportation and energy operators, in collaboration with the regional Civil Protection authorities, started working in 2009 to increase their knowledge of the existing interdependencies characterizing the Transportation and Energy critical systems for the Lombardy Region: on the basis of questionnaires, direct interviews and process mapping, operators were requested to provide information about critical infrastructures nodes, accident and service disruption events, maintenance, and crisis management internal processes and systems in order to understand fallout effects. A simulation of functional vulnerabilities and interdependencies has also been developed to support the programme activities (Trucco et al., 2011; Cagno et al., 2011). As a result of these cooperation efforts, the Lombardy Region Authorities have developed an emergency communication and information-sharing framework. The operators contributed to the identification of the relevant communication flows and channels under different emergency conditions and type of events. For example, in the scenario of heavy snowfall, the operators could state the kind of information they needed (e.g. very precise meteorological predictions), the information they could provide (particularly information that could have an impact for the other operators) and the role they expected the regional crisis management centre to play (e.g. to communicate with the public).

Integration of Citizens into Resilience Building

A third important point that needs to be considered in the context of CI resilience building is the incorporation of civil society. In this regard, citizens should be understood to be consumers and tax-payers, and thus stakeholders. It is therefore fundamental to communicate the CI decisions to be taken, particularly those decisions that concern the communities.

Since CI resilience should focus on maintaining the provision of certain infrastructure services, it is important to have information about specific dependencies (with respect to citizens but also regarding public facilities such as hospitals, transport systems, etc.). At the same time, it is equally important to view citizens as active contributors towards civil protection. For example, the principle underlying the Australian approach to CIR is that "communities are the heart of the resilience process" (Duckworth, 2012). For people to prepare for and respond to emergencies, people must understand the risks they live with. When they understand the risks, they need to be empowered to take action to deal with them. Governments cannot make people resilient, but they can help by providing information and ongoing support (e.g. Cabinet Office, 2013). Building resilience is a "bottom-up" process and governments as well as regional authorities need to agree on a communication framework to inform and engage people about risk. Although everyone must take into account the possibility of deficiencies in deliveries of services on which they are critically dependent, this is one area in which the State should assume greater responsibility (Duckworth, 2012). Accordingly, analysis of how top-down and bottom-up approaches can be matched is required.

Community resilience including the bottom-up approach is not a new concept, and has in fact been widely applied in the DRM community (Magis, 2010; Cutter et al., 2008; Paton & Johnston, 2000, 2006; Tobin, 1999), for example with respect to empowerment, education, access to institutions and resources (Edwards, 2009). With respect to the failure of CIs, however, the concept has only recently been applied.

Citizens can not only improve their own resilience by these measures but also serve as a source of information for civil protection agencies. People can provide relevant information to allow first responders to identify what is happening, where the priorities are, and what types of resources need to be mobilized. This may entail, for instance, analyzing information circulating among social networks, although this raises the question of validating and guaranteeing the accuracy of the information (e.g. Australian Emergency Management Institute, 2013).

CONCLUSIONS

The term 'resilience' is increasingly used for CI-related DRM strategies. However, it is applied in a variety of contexts and scales, often without a clear and stringent definition. This results in confusion around its meaning, so that it becomes difficult to understand what is meant when a resilience strategy is presented. This failure to clearly define the concept may mean that the actions and activities deriving from it fail to increase resilience. With respect to society and CIs, resilience strategies need to integrate the potential failure of infrastructure services instead of focussing only on their robustness and reliability. Relating resilience to concepts used in the DRM community and specifically to aspects of socio-ecological resilience facilitates the interrelation of technical systems while taking into account unexpected events— at least from a theoretical point of view. In order to operationalize a resilience framework for CIs towards natural hazards, further research is required. Although some current research projects address the general question of conceptualizing resilience in different contexts (e.g. the FP7 project EmBrace) and some valuable examples have been presented, for instance by TISP (2006), the operationalization potential for the CI context remains vague.

The implementation of resilience strategies by concrete measures and monitoring activities remains a challenge. In this regard, indicators of the efficiency of resilience strategies would be needed to evaluate results and to benchmark different approaches in order to generate arguments in favor of appropriate action. Difficulties in collecting and updating relevant data and information (Trucco et al., 2011, Robert & Morabito, 2010) are an obstacle to progress in this area. Technical solutions establishing a trustable and secure environment to exchange data and other information among different operators such as policy makers and operators canthereby facilitate the implementation of resilience strategies. Some first attempts have been made by Bruneau et al. (2003).

Finally, additional research with respect to bridging different temporal and spatial scales in resilience strategies is required. While operators consider operational business timeframes, governments and political decision makers follow a political and electoral timeframe. These timeframes differ from the lifespan of infrastructure systems or the timeframe of technical and technological changes. The superposition of these timeframes requires further analysis, while different fields— such as land use planning, emergency management or infrastructure development—take place at different spatial scales, which also have to be integrated.

REFERENCES

1. Abbate, J. (1999). From control to coordination: new governance models for information networks and other large technical systems. In: Coutard, O. (Ed.)*The Governance of Large Technical Systems*, pp. 114-129. London: Routledge.

2. Adger, W.N., T.P. Hughes, C. Folke, S.R. Carpenter & J. Rockström (2005). Social-ecological resilience to coastal disasters. *Science* 309: 1036-1039.

3. Aghababian R., C.P. Lewis, L. Gans & F.J. Curley (1994). Disasters within hospitals. *Annals of Emergency Medicine* 23: 771-777.

4. Alexander, D. (2000). *Confronting Catastrophe. New Perspectives on Natural Disasters*. Edinburgh: Dunedin Academic Press..

5. Australian Emergency Management Institute (2013). *National strategy for disaster resilience: community engagement framework, Handbook 6.*Commonwealth of Australia, Attorney General's Department. Archived at: http://www.webcitation.org/6Q3f6rtlk.

6.   Australian Government (2010). *Critical infrastructure resilience strategy*. Archived at: http://www.webcitation.org/6Q3g53aFs

7.   Balsells, M., B. Barroca, J. Amdal, Y. Diab, V. Becue & D. Serre (2013). Analysing urban resilience through alternative stormwater management options: application of the conceptual Spatial Decision Support System model at the neighbourhood scale. *Water Science and Technology* 68(11): 2448-2457.

8.   Berkes, F., J. Colding & C. Folke (Eds.) (2008). *Navigating Social-Ecological Systems: Building Resilience for Complexity and Change*. Cambridge: Cambridge University Press.

9.   Birkmann, J. (2013). Measuring vulnerability to promote disaster resilient societies: Conceptual frameworks and definitions. In: Birkmann, J. (Ed.)*Measuring Vulnerability to Natural Hazards: Towards Disaster Resilient Societies*, 2nd Edition, pp. 9-79. Tokyo: United Nations University Press.

10.   Blaikie, P., T. Cannon, I. Davis & B. Wisner (1994). *At Risk: Natural Hazards, People's Vulnerability, and Disasters*. London: Routledge.

11.   Boin, A. & A. McConnell (2007). Preparing for critical infrastructure breakdowns: the limits of crisis management and the need for resilience. *Journal of Contingencies and Crisis Management* 15(1): 50-59.

12.   Bompard, E., R. Napoli & F. Xue (2009). Analysis of structural vulnerabilities in power transmission grids. *International Journal of Critical Infrastructure Protection* 2(1–2): 5-12.

13.   Bouchon, S. (2006). *The vulnerability of interdependent critical infrastructures systems: epistemological and conceptual state of the art*. EUR 22205 EN. Ispra: Institute for the Protection and Security of the Citizen/European Commission.

14.   Bouchon, S. (2011). *Critical Infrastructures Identification: Reflexion about the European case*. PhD Thesis, Université de Nanterre Paris-X, France.

15.   Bouchon, S. & C. Dimauro (2012). Resilience: insights into the role of critical infrastructures disaster mitigation strategies. *Journal of Land Use, Mobility and Environment* 5(3): 103-117.

16.   Bouchon, S., C. Dimauro & P. Trucco (Eds.) (2012). *Proceedings of the 1stInternational Workshop on Regional Critical Infrastructure*

Protection Programmes: Main issues, Experiences and Challenges, Milan, 17-18 November 2011. Milan: Lombardy Region. Archived at:http://www.webcitation.org/6QPKwA22z.

17. Bruneau, M., S.E. Chang, R.T. Eguchi, G.C. Lee *et al.* (2003). A framework to quantitatively assess and enhance the seismic resilience of communities.*Earthquake Spectra* 19(4): 733-752.

18. Brunner, E.M. & M Suter (2008). *An inventory of 25 national and 7 international critical information infrastructure protection policies.* International CIIP Handbook 2008/2009. Zurich: Center for Security Studies, ETH Zurich.

19. Bunn, M. & O. Heinonen (2011). Preventing the next Fukushima. *Science* 333: 1580-1581.

20. Cabinet Office (2010). Strategic framework and policy statement on improving the resilience of critical infrastructure to disruption from natural hazards. March 2010. London. Archived at:http://www.webcitation.org/6Q2U9cifz

21. Cabinet Office (2013). *The role of local resilience forums: a reference document. The Civil Contingencies Act (2004), its associated Regulations (2005) and guidance, the National Resilience Capabilities Programme and emergency response and recovery.* Civil Contingencies Secretariat, July 2013 (V2). London. Archived at: http://www.webcitation.org/6Q3gBXBuR

22. Cagno, E., M. De Ambroggi & P. Trucco (2011). Interdependency analysis of CIs in real scenarios. In: Berenguer, C., A. Grall & C. Guedes Soares (Eds.) *Advances in Safety, Reliability and Risk Management: ESREL 2011* , pp. 2508-2514. London: CRC Press.

23. Cardona, O.D. (2011). Disaster risk and vulnerability: notions and measurement of human and environmental insecurity. In: Brauch, H.G., U. Oswald Spring, C. Mesjasz, J. Grin *et al.* (Eds.) *Coping with Global Environmental Change, Disasters and Security: Threats, Challenges, Vulnerabilities and Risks* pp. 107-122.

24. Cardona, O.D. & A.H. Barbat (2000). *El Riesgo Sísmico y su Prevención. [Earthquake risk and prevention.]* Cuaderno Técnico no.5. Madrid: Schneider Electric.

25. City of Cape Town (2006). Framework for adaptation to climate change in the city of Cape Town (FAC[4]T). Report submitted to

City of Cape Town: Environment Resource Management, August 2006. .Archived at:http://www.webcitation.org/6Q3ggPs1j

26.   City of New York (2013). *A stronger, more resilient New York*. A PlaNYC report. URL: http://s-media.nyc.gov/agencies/sirr/SIRR_singles_Lo_res.pdf.

27.   Cutter, S.L., L. Barnes, M. Berry, C. Burton *et al.* (2008). A place-based model for understanding community resilience to natural disasters. *Global Environmental Change* 18: 598-606.

28.   De Bruijne, M. & M. van Eeten (2007). Systems that should have failed: Critical Infrastructure Protection in an institutionally fragmented environment. *Journal of Contingencies and Crisis Management* 15: 18-29.

29.   DHS [U.S. Department of Homeland Security] (2009). *National Infrastructure Protection Plan. Partnering to enhance protection and resiliency.* Archived by WebCite® at: http://www.webcitation.org/6QRhJzY8f

30.   DHS [U.S. Department of Home Affairs] & Public Safety Canada (2010). *Canada-United States Action Plan for Critical Infrastructure.* Archived at: http://www.webcitation.org/6QRhFDyes

31.   Duckworth, M. (2012). *The resilience journey: the importance of people and communities at times of disaster.* Keynote speech at the 2nd International Workshop on Regional Critical Infrastructure Resilience, 15 November 2012, Edinburgh, UK.

32.   EC [European Commission] (2004). *Communication from the Commission to the Council and the European Parliament: Critical Infrastructure Protection in the fight against terrorism.* COM(2004) 702 final. Brussels, 20th October 2004. Archived at: http://www.webcitation.org/6QRhOqN7Q

33.   EC [European Commission] (2008). Council Directive 2008/114/EC of 8 December 2008 on the identification and designation of European critical infrastructures and the assessment of the need to improve their protection.*Official Journal of the European Union* L 345: 75-82.

34.   EC [European Commission] (2009). Principles of multi-risk assessment: interactions amongst natural and man-induced risks. Project report EUR23615. Brussels: EC. Archived at:http://www.webcitation.org/6Q2OpmqPY

35. Edwards, C. (2009). *Resilient Nation*. London: DEMOS. Archived at: http://www.webcitation.org/6Q3hRrFQ1

36. Fekete, A. (2012). Safety and security target levels: opportunities and challenges for risk management and risk communication. *International Journal of Disaster Risk Reduction* 2: 67-76.

37. Fekete, A., P. Lauwe & W. Geier (2012). Risk management goals and identification of critical infrastructures. *International Journal of Critical Infrastructures* 8(4): 336-353.

38. FMIG [Federal Ministry of the Interior of Germany] (2009). National strategy forcritical protection *(CIP Strategy)*. Berlin: FMIG. Archived at:http://www.webcitation.org/6Q2Titf7U

39. Folke, C. (2006). Resilience: the emergence of a perspective for social-ecological systems analyses. *Global Environmental Change* 16(3): 253-267.

40. Folke, C., S. Carpenter, T. Elmqvist, L. Gunderson *et al.* (2002). *Resilience and sustainable development: building adaptive capacity in a world of transformations*. Scientific Background Paper on Resilience for the process of The World Summit on Sustainable Development on behalf of The Environmental Advisory Council to the Swedish Government. April 2002. Stockholm: Environmental Advisory Council.

41. Füssel, H.-M. & R.J.T. Klein (2006). Climate change vulnerability assessments: an evolution of conceptual thinking. *Climatic Change* 75: 301-329.

42. Gallopín, G.C. (2006). Linkages between vulnerability, resilience, and adaptive capacity. *Global Environmental Change* 16: 293-303.

43. Garmezy, N., A.S. Masten & A. Tellegen (1984). The study of stress and competence in children: a building block for developmental psychopathology.*Child Development* 55: 97-111.

44. German Federal Government (2008). *Deutsche Anpassungsstrategie an den Klimawandel. [German strategy for adaptation to climate change.]* Decided by the Federal Cabinet on December 17, 2008 Archived at:http://www.webcitation.org/6Q3haRrRU

45. Gheorghe, A.V. & D.V. Vamanu (2005). Reading vulnerability in phase portraits: an exercise in probabilistic resilience assessment. *International Journal of Critical Infrastructures* 1(4): 312-329.

46.   Gheorghe, A.V. & D.V. Vamanu (2008). Quantitative vulnerability assessment of Critical Infrastructures: watching for hidden faults. *International Journal of Critical Infrastructures* 4(1/2): 144-152.

47.   Gheorghe, A.V., M. Masera, M. Weijnen & L. De Vries (2006). *Critical Infrastructure at Risk. Securing the European Electric Power System.* Dordrecht: Springer.

48.   Government of Finland (2006). *The strategy for securing the functions vital to society.* Government Resolution 23.11.2006, prepared by The Security and Defence Committee. Archived at:http://www.webcitation.org/6Q3hh1csj

49.   Greenberg, M., N. Mantell, M. Lahr, F. Felder & R. Zimmerman (2007). Short and intermediate economic impacts of a terrorist-initiated loss of electric power: case study of New Jersey. *Energy Policy* 35: 722-733.

50.   Gunderson, L.H. & C.S. Holling (Eds.) (2002). *Panarchy: Understanding Transformations in Human and Natural Systems.* Washington DC: Island Press.

51.   Hartong, M., R. Goel & D. Wijesekera (2008). Security and the US rail infrastructure. *International Journal of Critical Infrastructure Protection* 1: 15-28.

52.   Hellström, T. (2007). Critical infrastructure and systemic vulnerability: towards a planning framework. *Safety Science* 45: 415-430.

53.   Hémond, Y. & B. Robert (2012). Evaluation of state of resilience for a critical infrastructure in a context of interdependencies. *International Journal of Critical Infrastructures* 8(2/3): 1-18.

54.   Her Majesty the Queen in Right of Canada (2009). *Action plan for critical infrastructure.* Archived by WebCite® at: http://www.webcitation.org/6QRhtYqx3.

55.   HM Government [Her Majesty's Government] (2010). *A strong Britain in an age of uncertainty: the national security strategy.* Presented to Parliament by the Prime Minister by Command of Her Majesty. London: The Stationery Office Limited. Archived by WebCite® at: http://www.webcitation.org/6QRhxalGZ.

56.   Holling, C.S. (1973). Resilience and stability of ecological systems. *Annual Review of Ecology and Systematics* 4: 1-23.

57. Holling, C.S. (1996). Engineering resilience versus ecological resilience. In: Schulze, P. (Ed.). *Engineering Within Ecological Constraints*, pp. 31-44. Washington DC: The National Academies Press.

58. Holling, C.S. (2001). Understanding the complexity of economic, ecological, and social systems. *Ecosystems* 4: 390-405.

59. IRGC [International Risk Governance Council] (2006). *Managing and reducing social vulnerabilities from coupled critical Infrastructures*. White paper. Geneva: IRGC.

60. IRGC [International Risk Governance Council] (2010). *Emerging risks: sources, drivers and governance issues*. Concept note. Revised version, March 2010. Geneva: IRGC.

61. Klein, K.R., M.S. Rosenthal & H.A. Klausner (2005). Blackout 2003: preparedness and lessons learned from the perspectives of four hospitals.*Prehospital and Disaster Medicine* 20(5): 343-349.

62. Klein, R.J.T., R.J. Nicholls & F. Thomalla (2003). Resilience to natural hazards: how useful is this concept? *Global Environmental Change Part B: Environmental Hazards* 5(1–2): 35-45.

63. Kröger, W. (2008). Critical infrastructures at risk: a need for a new conceptual approach and extended analytical tools. *Reliability Engineering and System Safety* 93: 1781-1787.

64. Kröger, W. & E. Zio (2011). *Vulnerable Systems*. London: Springer.

65. Lauwe, P. & C. Riegel (2008). Schutz Kritischer Infrastrukturen: Konzepte zur Versorgungssicherheit. [Critical infrastructure protection: concepts for security of supply.] *Informationen zur Raumentwicklung* 1/2: 113-125.

66. Lhomme, S., D. Serre, Y. Diab & R. Laganier (2013). Assessing the resilience of the urban networks: a preliminary step towards more flood resilient cities.*Natural Hazards and Earth System Sciences* 13: 221-230.

67. Li, Q., J. Sun & J. Fan (2012). Seismic vulnerability assessment through explicit consideration of uncertainties in structural capacities and structural demands.*International Journal of Structural Engineering* 3(1/2): 27-36.

68. Luiijf, E., H. Burger & M. Klaver (2003). Critical infrastructure protection in the Netherlands: a quick-scan. In: Gattiker, U.E. (Ed.) *EICAR 2002 Conference Best Paper Proceedings* (ISBN: 87-987271-2-5). Copenhagen: EICAR.

69. Magis, K. (2010). Community resilience: an indicator of social sustainability.*Society & Natural Resources: An International Journal* 23(5): 401-416.

70. Monstadt, J. (2008). Der räumliche Wandel der Stromversorgung und die Auswirkungen auf die Raum- und Infrastrukturplanung. [The spatial change of the power supply and the impact on the spatial and infrastructure planning.] In: Moss, T., M. Naumann & M. Wissen (Eds.). *Infrastrukturnetze und Raumentwicklung: Zwischen Universalisierung und Differenzierung [Infrastructure Networks and Spatial Development: Between Universalization and Differentiation]*, pp. 187-224. Munich: Oekom.

71. NIAC [National Infrastructure Advisory Council] (2009). Critical *infrastructureresilience: final report and recommendations.* September 8, 2009. Archived at: http://www.webcitation. org/6QRi1SkGQ.

72. Norwegian CIP Commission (2006). *Protection of critical infrastructures and critical societal functions in Norway.* Report NOU 2006:6 submitted to the Ministry of Justice and the Police by the government appointed commission for the protection of critical infrastructure on 5th of April 2006 Archived by WebCite® at: http://www.webcitation.org/6Q3huvEYC

73. Paton, D. & D. Johnston (2000). Disasters and communities: vulnerability, resilience and preparedness. *Disaster Prevention and Management* 10(4): 270 - 277.

74. Paton, D. & D. Johnston (Eds.) (2006). *Disaster Resilience: An Integrated Approach.* Springfield: Charles C. Thomas.

75. Pelling, M. (2011). *Adaptation to Climate Change: From Resilience to Transformation.* London: Routledge.

76. Perelman, L.J. (2006). *Shifting security paradigms. Toward resilience.* CIPP Working Paper 10-06. Arlington, VA: George Mason University. Archived at: http://www.webcitation. org/6Q2PYzIxz

77. Petit, F., W. Buehring, R. Whitfield, R. Fisher & M. Collins (2011). Protective measures and vulnerability indices for the Enhanced Critical Infrastructure Protection Programme. *International Journal of Critical Infrastructures* 7(3): 200-219.

78. PCCIP [President's Commission on Critical Infrastructure Protection] (1997).*Critical foundations: protecting America's infrastructure.* The Report of thePresident's Commission on Critical Infrastructure Protection, October 1997. Washington: PPCIP. Archived by at:http://www.webcitation.org/6Q3iBqPud

79. Pursiainen, C. (Ed.) (2007). *Towards a Baltic Sea region strategy in Critical Infrastructure Protection.* Nordregio Report 2007: 5. Stockholm: Nordegio (Nordic Center for Spatial Development). Archived by WebCite® at: http://www.webcitation.org/6Q3iG90hh

80. Reed, D.A., K.C. Kapur & R.D. Christie (2009). Methodology for assessing the resilience of networked infrastructure. *IEEE Systems Journal* 3(2): 174-180.

81. Reichenbach, G., H. Wolff, R. Göbel & S. Stokar von Neuforn (2008). *Risiken und Herausforderungen für die öffentliche Sicherheit in Deutschland: Szenarien und Leitfragen.* Grünbuch des Zukunftsforums Öffentliche Sicherheit. [*Risks and challenges for public security in Germany: scenarios and questions.* Green Paper of the future forum public safety.] Berlin: Zukunftsforums Öffentliche Sicherheit. Archived at: http://www.webcitation.org/6Q2UfYjeG

82. Rich, E., J.J. Gonzalez, Y. Qian, F.O. Sveen, J. Radianti & S. Hillen (2009). Emergent vulnerabilities in Integrated Operations: a proactive simulation study of economic risk. *International Journal of Critical Infrastructure Protection* 2(3): 110-123.

83. Rinaldi S.M., J.P. Peerenboom & T.K. Kelly (2001). Identifying, understanding and analyzing critical infrastructure interdependencies. *IEEE Control Systems Magazine* 21 (6): 12-25.

84. Robert, B. & L. Morabito (2010). An approach to identifying geographic interdependencies among critical infrastructures. *International Journal of Critical Infrastructures* 6(1): 17-30.

85. Rutter, M. (1985). Resilience in the face of adversity. Protective factors and resistance to psychiatric disorder. *The British Journal of Psychiatry* 147: 598-611.

86. Scottish Government (2011). *Secure and resilient: a strategic framework for critical national infrastructure in Scotland.* Edinburgh: The Scottish Government. Archived by WebCite® at: http://www.webcitation.org/6Q3iPsVtR

87. Serre, D., B. Barroca & R. Laganier (2013). *Resilience and Urban Risk Management.* New York: CRC Press.

88. Smit, B. & J. Wandel (2006). Adaptation, adaptive capacity and vulnerability.*Global Environmental Change* 16(3): 282-292.

89. Suter, M. (2011, November). *PPP models for CIP implementation at regional level.* Paper presented at the *1st International Workshop on Regional Critical Infrastructure Protection Programmes: Main Issues, Experiences and Challenges,* 17-18 November 2011, Milan, Italy.

90. Svensson, G. (2008). Mutual and interactive vulnerability in supply-chain dyads.*International Journal of Logistics Economics and Globalisation* 1(2): 123-140.

91. The White House (2013). *Critical infrastructure security and resilience.*Presidential Policy Directive/PPD-21, February 12, 2013. Archived at: http://www.webcitation.org/6Q3iVRiUP

92. Thomassen, E. (2012, November). *Critical Infrastructure Resilience: the Norwegian approach.* Paper presented at the 2nd International Workshop on Regional Critical Infrastructure Resilience, 15 November 2012, Edinburgh UK.

93. Tierney, K. & M. Bruneau (2007). Conceptualizing and measuring resilience. A key to disaster loss reduction. *TR News* 250(May-June): 14-17.

94. TISP [The Infrastructure Security Partnership] (2006). *Regional Disaster Resilience: A Guide for Developing an Action Plan.* Reston, Virginia: ASCE (The American Society of Civil Engineers. Archived at: http://www.webcitation.org/6Q3ia2Lo4

95. TNO [Toegepast Natuurwetenschappelijk Onderzoek] (2011). *RECIPE [Recommended Elements of Critical Infrastructure Protection]: Good Practices Manual for Policy Makers in Europe.* TNO. Archived by WebCite® at:http://www.webcitation. org/6Q3ieefLF

96. Tobin, G.A. (1999). Sustainability and community resilience: the holy grail of hazards planning? *Global Environmental Change Part B: Environmental Hazards*1(1): 13-25.

97. Toubin, M., R. Laganier, S. Gomez, Y. Diab & D. Serre (2014). Improving the conditions for urban resilience through interdependencies identification and collaborative learning

between Parisian urban services. *Journal of Urban Planning and Development, ASCE* (in press).

98. Trucco, P., E. Cagno & M. De Ambroggi (2011). Dynamic functional modelling of vulnerability and interoperability of Critical Infrastructures. *Reliability Engineering and System Safety* 105: 51-63.

99. Tyler, S. & M. Moench (2012). A framework for urban climate resilience. *Climate and Development* 4(4): 311-326.

100. UNISDR [International Strategy for Disaster Reduction] (2007). *Hyogo Framework for Action 2005-2015: building the resilience of nations and communities to disasters.* Extract from the final report of the World Conference on Disaster Reduction (A/CONF.206/6). Geneva: United Nations. URL:http://www.ISDR.org/files/1037_hyogoframeworkforactionenglish.pdf

101. Vugrin, E.D., D.E. Warren, M.A. Ehlen & R.C. Camphouse (2010). A framework for assessing the resilience of infrastructure and economic systems. In: Gopalakrishnan, K. & S. Peeta (Eds.) *Sustainable and Resilient Critical Infrastructure Systems*, pp. 77-116.

102. Walker, B.H., C.S. Holling S.R. Carpenter & A.P. Kinzig (2004). Resilience, adaptability and transformability in social-ecological systems. *Ecology and Society* 9(2): 5.

103. Wise, C.R. (2006). Organizing for homeland security after Katrina: is adaptive management what's missing? *Public Administration Review* May/June: 302-318.

Impacts of Natural Disasters on Environmental and Socio-Economic Systems: What Makes the Difference?

Herlander Mata-Lima[I], Andreilcy Alvino-Borba[II], Adilson Pinheiro[III], Abel Mata-Lima[IV], and José António Almeida[V]

[I]Investigador Integrado do CERENA - Centro de Recursos Naturais e AmbientResearcher at CERENA - Centre for the Environment and Natural Resources,Instituto Superior Técnico da Universidade Técnica de Lisboa. Higher Technical Institute at the Technical University of Lisbon. Professor Visitante do Centro de Engenharia, Modelagem e Ciências Sociais Aplicadas (CECS), Universidade Federal do ABC (UFABC), São Paulo, BrasVisiting Professor at the Centre of Engineering, Modelling and Applied Social Sciences (CESC), Federal University of ABC (UFABC), São Paulo, Brazil

[II]Calaboradora do CERENA - Centro de Recursos Naturais e Ambiente, Instituto Superior Técnico da Universidade Técnica de Lisboa, Lisboa (IST/UTL), Portugal. Bolsista do CNPq, BrasiMember of staff at CERENA - Centre for the Environment and Natural Resources, Higher Technical

Institute at the Technical University of Lisbon, Lisbon (IST/UTL), Portugal. CNPq scholarship, Brazil

[III]Fundação Universidade Regional de Blumenau (FURB), Santa Catarina, BrasiBlumenau Regional University Foundation (FURB), Santa Catarina, Brazil

[IV]CTB, Universidade Politécnica de Madrid, Madrid (Spain).esCTB, Madrid Polytechnic University (Spain)

[V]CICEGe - Centro de Investigação em Ciência e Engenharia Geológica, Faculdade de Ciências e Tecnologia da Universidade Nova de Lisboa (FCT/UNL), Portugal.CICEGe - Centre of Research in Geological Science and Engineering, Faculty of Sciences and Technology of the New University of Lisbon (FCT/UNL), Portugal

ABSTRACT

This study addresses the environmental and socioeconomic impacts of natural disasters and focuses on the factors that can contribute to reducing damage both in material terms and in terms of loss of human life. A reflective analysis was carried out - based on a qualitative and quantitative approach - integrating environmental, economic and social dimensions of sustainability as well hydro-meteorological, climatological and geophysical paradigms of disasters (Hazard-Risk-Vulnerability-Resilience). Our objective is to identify key variables in the reduction of vulnerability and the prevention and mitigation of the impacts of natural disasters. The results stress that social capital, related to social and economic structures, exerts a significant influence as a factor which reduces the vulnerability of affected communities.

INTRODUCTION

Natural disasters are caused by hydro-meteorological, climatological, geophysical and biological phenomena which adversely impact on the natural and built environment of affected regions. Their effects in terms of victims and material damage exceed the capacity for self-recovery of local communities, making external assistance necessary (vide GUHA-SAPIR et al., 2012; NOY, 2010; ALCÁNTARA-AYALA, 2002, p. 109-110).

The World Bank & United Nations report (2010) states that disasters expose the cumulative effects of decisions (individual and collective) previously taken in terms of land management (including unregulated growth of urban areas), construction techniques, implementation of sanitation infrastructure and low investment in educational programs, poverty reduction and social integration, among others. Such decisions combined with high intensity natural events (e.g. floods, landslides, storms and earthquakes) provoke an array of socioeconomic and environmental impacts.

A trans-disciplinary approach to the underlying concept of natural disasters suggests that they are characterized by naturally occurring events whose consequences are often aggravated by man-made actions which surpass the capacity of man's built infrastructure to contain. They result in tragic disturbances in the social and environmental sphere together with socioeconomic impacts of extreme severity, such as high levels of material damage, the loss of life and means of subsistence for affected communities, and the spread of infectious diseases[i] due to the degradation of sanitary conditions. They are consequently responsible for a series of adverse environmental and socio-economic impacts due to the way they cause disturbances (or imbalances) in the *environmental* (CHINO et al., 2011; McENTIRE, 2001; ADRIANTO & MATSUDA, 2002), *economic* (DAVIS et al., 2012; FREITAS et al., 2012; LOAYZA et al 2012; NOY & VU, 2010; UN, 1999) and social (GUHA-SAPIR et al., 2012; TAKAHASHI et al., 2012; O'BRIEN et al., 2006; YODMANI, 2001) aspects of sustainability.

[i]Concerning infectious diseases TAKAHASHI *et al.* (2012) emphasise the fact that the affected community is exposed to infectious contamination agents during the initial post-disaster phases, such as rescue and recovery in provisional camps.

In the last two decades many studies have consistently presented forecasts and demonstrations of an increase in the frequency and intensity of natural disasters (e.g. hurricanes, floods, droughts and associated forest fires, earthquakes, tornadoes, among others), above all those related to climate factors (*vide* GUHA-SAPIR et al., 2012; IPCC, 2007; VINK et al., 1998) and the relation between natural disasters and the macro-economic indicators of different countries (SCHUMACHER & STROBL, 2011; LOAYZA et al. 2012; NOY, 2010).

This issue has taken on particular importance as the Intergovernmental Panel on Climate Change (IPCC, 2007) report states that one of the consequences of global warming is the likely increase in the frequency and intensity of extreme climatic events (above all in tropical regions), which together with disasters caused by geophysical factors (e.g. earthquakes, tsunamis, volcanic eruptions) comprise a strong threat to developing countries (NAUDE, 2010; IFRC, 2003, 2010; O'BRIEN *et al.*, 2006). As is well known, these countries have low resilience in face of disasters (EBEKES & COMBES, 2013; CUARESMA, 2010; WORLD BANK & UNITED NATIONS, 2010).

Natural disasters, even when they are classified as small or moderate (DATAR *et al.*, 2013), are responsible for adverse socio-economic and environmental impacts (GUHA-SAPIR *et al.*, 2012), particularly in underdeveloped regions (or regions in development) (TOYA & SKIDMORE, 2007; WORLD BANK & UNITED NATIONS, 2010). This is due to both a lack of preventive action plans and resources and to low resilience, inherent to low levels of social capital[ii] (*vide* TOYA & SKIDMORE, 2007, p. 20-21; JACOBI & MONTEIRO, 2006, p. 27; ALCÁNTARA-AYALA, 2002, p. 108), which contribute to the prolongation of the adverse effects on the environment and society. This prolonged duration causes a greater spatial dispersal of environmental impacts where natural agents (e.g. water, wind) transport the problem beyond its source and aggravate socio-economic impacts by disturbing economic activity (e.g. agriculture, trade, tourism) and increasing social vulnerability.

[ii]Social capital is the result of structural characteristics of social organization which encourage the formation of networks, standards, value systems, relations of trust and participative engagement so as to facilitate coordination and cooperation for the common good. (*vide*, e.g.PARK *et al.*, 2012, p. 1512).

As an example of the influence of social capital it is worth emphasizing Alcántara-Ayala (2002, p.108) who argues that one of the causes of natural disasters in poor or developing countries is:

...related to the historical development of these countries, where the economic, social, political and cultural conditions are poor and consequently lead to increased vulnerability to natural disasters (economic, social, political and cultural vulnerability) [our translation].

This paper addresses natural disasters whose origin and scale are not limited to natural causes, in other words where the causes and the effects are also closely related to demographic and industrial growth, something inherent to the socio-economic growth of contemporary societies. The industrial and demographic growth, which encompasses the combined effects of population in a biological sense and the effects of production-consumption in a technological sense (ALVINO-BORBA & MATA-LIMA, 2011; WETZEL, 1996), is normally associated to an increase in density whether in terms of population or infrastructure (built environment), where both factors have aspects and impacts (environmental and socio-economic) which contribute to an increase in the scale of natural disasters and to the worsening of vulnerabilities of affected communities.

It is important to stress that in accordance to the ISO 14001 norm: (i) *environmental aspect* is the element of an organization's activities, products and services which may interact with the environment; while (ii) *environmental impact* is any change to the environment, adverse or beneficial, which is a result, fully or partly, of environmental aspects of the organization.

In this context, the environmental aspect is related to the cause of the problem or to an environmental improvement, while the environmental impact is related to the effect of the problem or to an environmental improvement. Therefore, environmental aspects should be identified based on the following factors (*vide*, e.g., MARAZZA *et al.*2010; UNIVERSITY OF STRATHCLYDE, 2000): (i) social inclusion; (ii) economic development; (iii) use of resources; (iv) transport; (v) environmental and ecological protection.

The aspects addressed above are a list of variables which must be considered in the production of development programs and the implementation of disaster prevention plans. Sustainable development, as is well known, must address environmental, social and economic aspects in a transversal and balanced way, always using the best available technology to achieve stated objectives, as presented in Figure 1.

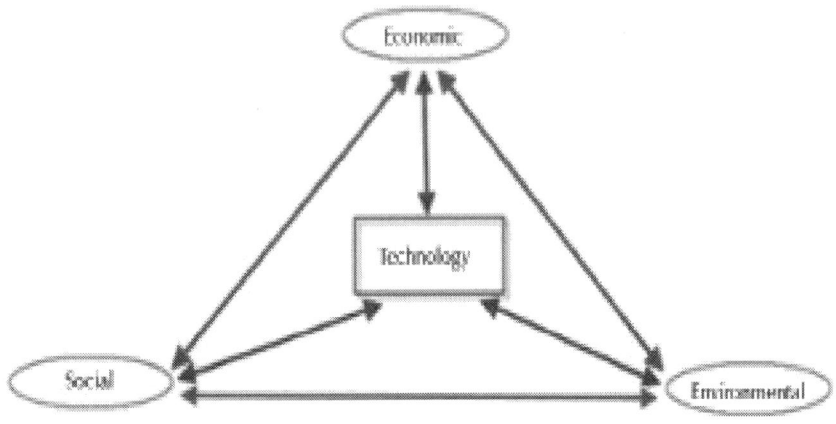

Figure 1: Dimensions of the sustainability triangle.

The sustainability triangle allows us to leave aside many considerations which have been widely addressed in previously published studies, such as that of MAUERHOFER (2008, p. 498).

NATURAL DISASTERS

Origin and Occurrence

Natural disasters are generally classified as having hydrological, meteorological, climatic, geophysical or biological causes/origins (GUHA-SAPIR *et al.*, 2012). In this paper natural disasters caused by hydrological and meteorological phenomena will be grouped in one category denominated hydro-meteorologic, and will not include disasters with a biological origin (these are less common), as presented in *Table 1*.

Table 1: Main natural disasters caused by hydro-meteorologic, climatic and geophysical phenomena

Disasters			Relevant observations
Origin	Hydro-meteorologic	Hurricane	Most frequent natural disasters, accounting for 77.4% of the total in 2011(GUHASAPIR et al., 2012, p.2); floods is the category which has caused most deaths in history and Brazil stood out globally in 2011 with 900 deaths. Hydro-meteorological disasters cause the most concern for Small Island Development States (SIDS), and also Small Islands Economies (SIE) which are part of an archipelago (e.g. Japan) (cf. UN, 1999)
		Floods	
		Tornado	They occur in all continents, but predominate in Africa and the Americas, including Brazil, according to NEDEL (2012, p.120)
	Climatologic	Drought	These types of event occur from time to time throughout the world though, with the exception of some sub-Saharan countries (e.g. Ethiopia, Somalia and Kenya), they result in fewer victims (GUHA-SAPIR et al., 2012, p. 15). According to the same authors, from 2001 to 2010 climatic disasters represent an average of 12.9% of all natural disasters. It is the only natural disaster that does not predominate in Asia; it is more common in Europe. However, in Europe and Australia there are fewer victims of climatic disasters.
		Fire	
		Extreme Temperatures	The main consequences of this type of event are: destruction of forests, increased susceptibility of land to erosion and degradation of surface waters due to transport of waste through surface run-off.
	Geophysic	Earthquake	Geophysical disasters were responsible for 69,098 deaths from 2001 to 2010 (GUHA-SAPIR et al., 2012, p.2). The same authors state that in 2011 geophysical disasters were responsible for 68.1% of total deaths caused by natural disasters. These disasters predominate in Asia.
		Tsunami	
		Volcanic Eruption	
		Mass Movements	

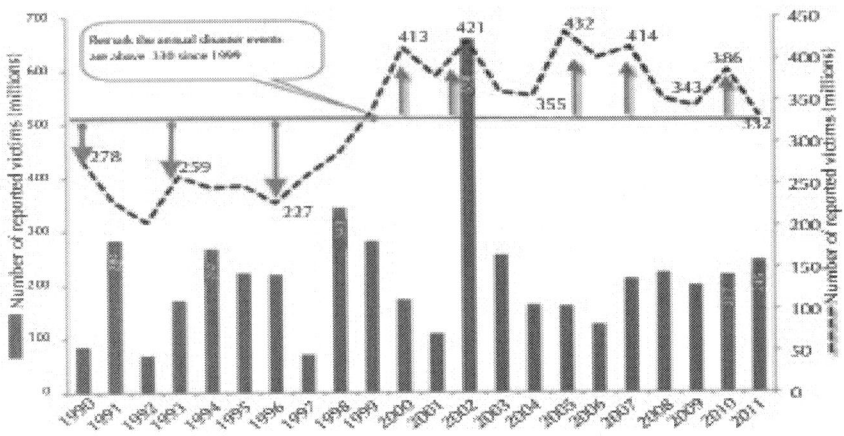

Figure 2: Evolution of occurrences of natural disasters and associated victims.

Figure 2 (modified from GUHA-SAPIR et al., 2012, p. 3) shows the global occurrence of natural disasters from 1990 to 2011 and their respective victims.

The approach taken in terms of addressing natural disasters is separated into four (4) disaster paradigms (cf.FRERKS et al., 2011, p. 106): Hazard-Risk-Vulnerability-Resilience. Table 2 is a descriptive summary of these paradigms where a distinction is made for those disasters where, in terms of intervention plans, an effort is made to reduce (↓) and increase (↑).}

Table 2: Description of disaster paradigms

Paradigms	Description
Hazards (↓)	The probability of a potentially damaging natural phenomenon occurring in a specific place and in a specific period of time (TOMINAGA et al., 2009, p. 151). REBELO (2008) presents a comprehensive explanation of the concepts of hazard and risk.

Risk ($\downarrow$)	A combination of the probability of an event occurring and its severity (negative consequences) TOMINAGA et al. (2009, p. 149), frequently expressed as a product of hazard in terms of its consequences for man.
Vulnerability ($\downarrow$)	The combination of processes and conditions which result from physical, social, economic and environmental factors, increasing the susceptibility of a community (exposed to the risk) to the impact of dangers (TOMINAGA et al., 2009, p. 151). Vulnerability refers to the capacity of a community to anticipate, confront, resist and recover from the impacts of natural disasters and it comprises a variety of factors which determine the degree of exposure of both people and material goods to risk (INGRAM et al., 2006, p. 607).
Resilience ($\uparrow$)	Resilience is defined as the capacity of a community to resist and recover from adversity, both in the short and long-term (NHHS, 2009 apud FRERKS et al., 2011, p. 112). However, the definitions in GIBBS (2009, p. 324) and KLEIN et al.(2003, p. 35) seem to be better suited to the reality in that they consider resilience to be a simple attribute related to the level of disturbance that a system can absorb without losing its capacity and ability to re-organize itself. Here, resilience is only considered as one of the factors which influence the adaptive capacity of the system.

Environmental and Socio-Economic Aspects of Disasters

Environmental Aspect

The environmental aspect (stricto sensu) of natural disasters has been widely addressed in the specialized technical bibliography (vide, e.g., SRINIVAS & NAKAGAWA, 2008, p. 6; AERTS & BOTZEN, 2011) and a summary is presented in Table 3.

This section aims to highlight the strong relationship of interdependence which exists between protection and conservation of bio-physical factors (e.g. land, water, atmosphere, fauna and flora) and socio-economic development. The growth in the development of rural tourism[iii] (HAVEN-TANG & JONES, 2012) which essentially exploits activities inherent to rural regions is an example which underlines this affirmation (HAVEN-TANG & JONES, 2012; SRINIVAS & NAKAGAWA, 2008). On the other hand, it is known that natural disasters are closely related to coastal zones (YASUHARA et al., 2012; COSTANZA & FARLEY, 2007), fundamental elements in providing a competitive advantage to seasonal summer tourism in developing countries (in Africa, Latin America and Asia).

iii Tourism should not only be interpreted as activities related to the agricultural sector, as it encompasses various activities, such as engaging with Nature (e.g. ornithology), adventure activities, sport, health (e.g. ethnomedicine), education, art and heritage (*vide*, e.g., SU, 2011, p. 1438).

Table 3 synthesizes the environmental aspects of a man-made nature which exacerbate natural disasters. The table highlights a number of conspicuous examples of environmental aspects (causes of impacts) connected to engineering mega-projects which are likely to cause large-scale population movements, among many other significant negative environmental impacts with a wide variety of consequences. These projects are usually supported by viability studies which point to the generation of multiple positive socio-economic externalities for the regions where they are implemented, such as economic growth resulting from the revitalization of existing activities, the creation of new investment opportunities and, above all, employment for the local population (*vide*, e.g., MATA-LIMA, 2009).

Table 3: Examples of the relationship between aspects of economic growth and natural disasters

Environmental Aspect	Relationship with natural disasters (environmental and socio-economic impacts)
Road networks	Frequently worsen the impacts of flash floods and landslides as the location of road networks in relation to the hydro-graphic network changes the balance between the intensity of the flood (and the flow of residues) and the resistance of water lines (including riverside zones) (vide JONES et at, 2000, p. 80). The destruction of roads during a disaster causes problems for the movement of people and goods between, for example, urban and rural zones.
Building in flood prone areas	Makes communities more vulnerable to floods, transforming a phenomenon, which in a situation where good land use and planning practices have been adopted would be less catastrophic, into a disaster with elevated levels of material damage and loss of life (AERTS & BOTZEN, 2011, p. 8). It is worth emphasising that more than half the world's population live in urban areas which has increased the density of the built environment, caused traffic chaos and, naturally, leading to heightened difficulties in evacuating in emergency situations
Dam reservoirs	In the case of earthquakes the water stored in reservoirs is launched downstream as the dam wall breaks, causing high levels of material damage and victims, as well as destroying lake and riverside zones. A disaster of this type occurred in the southeast of China (Sichuan) in 1786, causing more than 100,000 deaths (DAI et al., 2005, p. 205).
Nuclear power plants	Earthquakes and subsequent tsunamis may cause the destruction of nuclear plants, releasing radioactive substances into the environment (e.g. Fukushima Daiichi Nuclear Power Plant in 2011 — Japan) (vide CHINO et at, 2011), as well as spreading infectious diseases (TAKAHASHI, et al. 2012).

Oil exploration	Earthquakes may cause the collapse of oil-producing infrastructures (SKOGDALEN & VINNEM, 2012, p. 62) resulting in the release of oil into the sea or on land, depending on whether it is an offshore or onshore platform.

Table 3 helps to clarify the assertions made by other authors (TOYA & SKIDMORE, 2007, p. 20; ALCÁNTARA-AYALA, 2002, p. 108; YODMANI, 2001, p. 2) that natural disasters are not extreme phenomena exclusively caused by nature. Indeed, given that vulnerability is a determining factor in the impact of disasters it can be argued that the development model adopted by the human race also significantly contributes to disasters taking place.

Socio-Economic Aspect

The growth in socio-economic aspects of disasters has shown an increase (*vide Figure 3*) due to the direct impacts on vulnerable communities. These often conceal environmental impacts and therefore are deserving of special attention on the part of agents, politicians and researchers who are responsible for finding solutions to mitigate their effects.

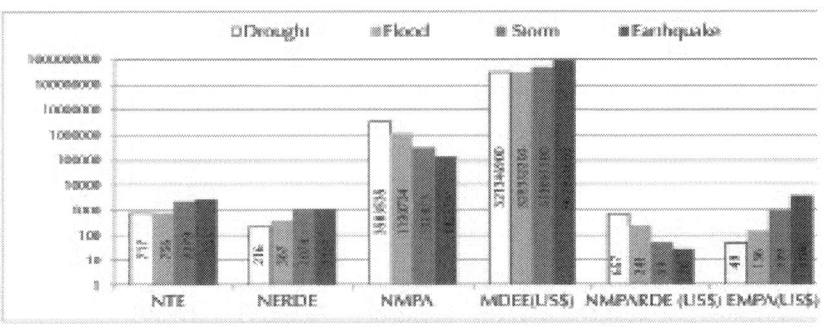

Figure 3: Social and economic costs of natural disasters from 1961 to 2005 (taken from the EM-DAT database and summarized by LOAYZA et al. 2012, p.1318).

Loayza et al. (2012, p. 1317) recently stressed that natural disasters cause significant economic and physical damage whose effects can spread beyond the immediate locality. They also observed that the impact of disasters on economic growth is not always negative and that developing countries are more vulnerable to these disasters as more sectors are affected. This is intrinsically related to the heightened degree of vulnerability and the low resistance of these countries. The WORLD BANK & UNITED NATIONS (2010) draws attention to the fact that in underdeveloped regions economic growth rarely occurs after natural disasters as the intensity of the negative effects depends on the structure of the economy. Moreover, it is known that regions with low social capital also have weak economic structures and experience difficulties in securing adequate resources to address the problems caused by disasters.

It is also important to account for the following peculiarities of socio-economic aspects:

Herlander Mata-Lima[I], Andreilcy Alvino-Borba[II], Adilson Pinheiro[III], Abel Mata-Lima[IV], and José António Almeida[V]

- Remittances significantly mitigate the impacts of natural disasters in terms of the number of victims in developing countries, accounting for between 8% and 17% of Gross National Product (GNP) (cf. EBEKE & COMBES, 2013);
- As natural disasters affect the poorest countries more than others, the most vulnerable and marginalized populations have to deal with the most serious consequences (FREITAS et al., 2012; IFRC, 2003, 2010).
- Table 4 is a good illustration of how the vulnerability of poor regions contributes to a significant increase in the negative impacts of natural disasters. Furthermore, based on data from the Center for Research on the Epidemiology of Disaster (CRED), globally there are more deaths from disasters and higher economic costs as time progresses, as O'BRIEN et al. (2006) emphasizes;

Table 4: Basic characteristics and consequences of earthquakes in Haiti and Japan

Country	General data (2010)	Year and basic characteristics	consequences: human victims and economic cost	Source
HAITI (poor country)	Population: 9,993,247 inhabitants GNP per capita (US$): 664 Annual growth of GNP per capita:7% Life expectancy at birth: 62 years	2010: Earthquake of 7.0 to 7.3 on the Richter scale, lasting 35 seconds	Approximately 230,000 deaths and 2 million people affected (15% of population). Economic cost equivalent to 120% of GNP	FREITAS et al. (2012)
JAPAN (developed country)	Population: 127,450,459 inhabitants GNP per capita (US$): 43 063 Annual growth of GNP per capita:5% Life expectancy at birth: 83 years	2011: Earthquake of 9.0 on the Richter scale, followed by a tsunami which caused water levels to rise 35 m	Approximately 19,000 deaths. Economic cost more than 5.4% of GNP	LIN et al. (2012); WORLD BANK*

* Available at: <http://databank.worldbank.org>. Accessed: July 2012.

- The increase in the number of disasters and their consequences is related to an increase in the vulnerability of communities throughout the world as a result of the development model adopted. The increase of vulnerability is not uniform, as there are significant variations between regions, nations, provinces, cities, communities, socio-economic classes, castes and even gender (cf. YODMANI, 2001);

- Urban areas benefit from having better physical infrastructure (e.g. hospitals, civil protection services, sanitation systems and other logistics) and administrative support systems (e.g. emergency plans); indeed, prevention and intervention plans are more likely to exist in urban areas (IFRC, 2010). However, the fact that the largest cities in the world are in poor and developing countries - such as São Paulo, whose problems are highlighted by JACOBI & MONTEIRO (2006, p. 32-33) and which is located in a country where hydro-meteorological disasters predominate - makes the scenario extremely worrying as these cities lack the above mentioned infrastructure.

MANAGEMENT OF ENVIRONMENTAL AND SOCIO-ECONOMIC IMPACTS ASSOCIATED TO NATURAL DISASTERS

In the previous sections we concentrated on establishing a relation between the environmental aspects and impacts of the most common natural disasters (e.g. floods, landslides), demonstrating the interdependence between the social, economic and environmental aspects of sustainability. This approach aims to make clear the complicit relationship between these three aspects of sustainability and the four disaster paradigms as a starting point in order to draw up and implement a management plan for preventing disasters. This effort is fundamental, as already mentioned, since reducing vulnerability depends on systematically tackling the complex interactions between inherent physical, environmental and social factors (*vide, e.g.,* INGRAM et al.2006).

Preventive Management

Though it is not humanly possible to adopt measures to eliminate the extreme phenomena which cause natural disasters, preventive planning is vital in mitigating impacts on socio-economic and environmental systems, particularly those which are the most vulnerable, as a way of increasing the degree of resilience of local communities. In this context it is worth stressing the words of McENTIRE (2001, p. 189): "The central argument to be made is that vulnerability is, or should be, the key concept for disaster scholarship and reduction". This concern reflects the final recommendation of the *World Summit on Sustainable Development* (WSSD) which emphasizes the need for an integrated approach to include vulnerability, risk evaluation and disaster management by focusing on the prevention and mitigation of impacts (UNISDR, 2003; WORLD BANK & UNITED NATIONS, 2010).

The management approach should be flexible and preventive, adopting the following stages. It is important to emphasize that often efficient preventive management may require cross-border cooperation (e.g. involving a number of countries) in cases where the scale and nature of the disaster demand it (e.g. floods in shared water basins, forest fires in border areas).

Identifying environmental aspects and impacts is fundamental in managing risks, and this should be the first step in a risk management study. This first stage is called *establishment of context* as Pojasek's flowchart shows (2008, p.97) in Figure 4.

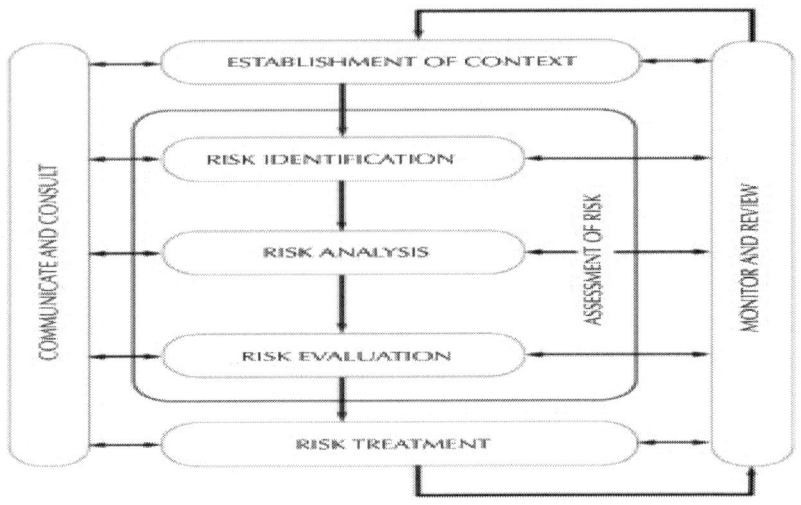

Figure 4: Procedures for Risk Management.

It is clear that *establishment of context* is of paramount importance in evaluating the degree of severity of impacts, in that these are more pronounced (and socially visible) when dealing with urban and populous regions where a considerable amount of infrastructure is built in risk zones, drastically affecting socio-economic aspects. As risk analysis is essentially based on the probability of a given event occurring and the degree of severity of the resulting consequences (*vide*, e.g., KORTENHAUS E KAISER, 2009; TOPUZ *et al.*, 2011), it is evident that the local bio-physical and socio-economic context must be assigned a determining role in the contextualization and evaluation of the risk.

SUMMARY AND RECOMMENDATIONS

The answer to the question contained in the title (what makes the difference?) can be found, above all, in social capital, as this has a determining influence as a factor of vulnerability given that the developed nations (e.g. Japan, USA) - despite having significantly fewer victims of natural disasters - are no less affected by extreme phenomena (e.g. hydro-meteorological) capable of provoking disasters than the poorest nations, as underlined by other authors (e.g. GUHA *et al.*, 2012; KAHN, 2005).

The following aspects which play a key role in the mitigation of natural disasters should be emphasized:

- Natural disasters should be approached from a trans-disciplinary perspective as their prevention and mitigation requires technical-scientific cooperation between different areas of science, engineering, economics, health, social studies and law. In addition, stakeholder participation (e.g. local community) is a *sine-qua-non* in reducing their socio-economic and environmental impacts.

- Vulnerability must be dealt with by increasing the social capital of communities which are located in regions of heightened risk of disasters. This can be achieved through education/training and by fostering citizenship which advocates participation in collective actions; reducing isolation by creating networks which encourage contact and exchange of experiences between different communities with concerns in common in terms of the risk management of disasters; among other actions aiming at building social capital.

- Natural disasters in developing countries cause impacts, particularly in terms of the degradation of health (DATAR *et al.*, 2013), due to diseases related to a worsening of environmental sanitation conditions, as Takahashi, *et al.* (2012) has emphasized;

- Globally, greater attention and more proactive intervention is necessary (in terms of prevention planning) on the part of governments and NGOs, as set out by the World Bank & United Nations (2010);

- There needs to be investment and natural disaster prevention subsidies as well as authorities and organizations who are directly responsible for preventing disasters, as this can significantly reduce the number of victims and extent of material damage;

- Lessons must be learnt from disasters and the post-disaster period should be an opportunity to implement good practices in terms of land use and integrating flexible measures instead of rushing to rebuild on a huge scale which, in some cases, may increase the vulnerability of local communities to future events.

Among aspects which help to mitigate disasters, social capital is fundamental in creating the conditions to reduce vulnerability, and consequently, the dependency of communities (or nations) on external initiatives.

This is because social capital is paramount in creating the necessary social, economic and political structures (including cooperation and inclusion in international networks) to foster socio-economic development based on an agreed path of sustainable development. This in turn contributes to a reduction of the level of risk communities are exposed to.

Furthermore, it is important to stress that an analysis of the spatial-temporal evolution of the data on disasters shows that nations which have a higher gross national product (GDP), a more educated population, more social and political freedom providing the conditions for effective and active citizenship, and a more comprehensive financial system suffer fewer losses when extreme phenomena occur which provoke natural disasters (*vide*, e.g., OXLEY, 2013; TOYA & SKIDMORE, 2007).

In terms of preventing natural disasters it is extremely important to create an appropriate context involving pro-active measures where community adaptation to climate changes and to reducing exposure to risk leads to both a reduction in vulnerability and, consequently, a reduction in the scale of the socio-economic impacts which are evident today in poverty-stricken regions where disasters occur.

ACKNOWLEDGEMENTS

O primeiro autor agradece ao CNPq pelo apoio concedido no âmbito do projeto "Geo-environmental modelling using strategic environmental assessment that incorporates biophysical factors and stakeholder engagement via transdisciplinary approach" (Processos 407507/2012-4 & 401425/2012-6) que estimulou a realização deste artigo.The first author would like to thank CNPq (Brazilian Council for Scientific and Technological Development) for their support for the "Geo-environmental modelling using strategic environmental assessment that incorporates biophysical factors and stakeholder engagement via transdisciplinary approach" project (Processes 407507/2012-4 & 401425/2012-6) which led to the writing of this article.

REFERENCES

1. ADRIANTO, L., MATSUDA, Y. Developing Economic Vulnerability Indices of Environmental Disasters in Small Island Regions. Environmental Impact Assessment Review, v. 22, p. 393-414, 2002.

2. AERTS, J., BOTZEN, W. Flood-resilient waterfront development in New York City: Bridging flood insurance, building codes, and flood zoning. Ann. N.Y. Acad. Sci., v.1227, p.1-80, 2011.

3. ALCÁNTARA-AYALA, I. Geomorphology, Natural Hazards, Vulnerability and Prevention of Natural Disasters in Developing Countries. Geomorphology, v. 47, n. 2-4, p. 107-124, 2002.

4. ALVINO-BORBA, A., MATA-LIMA, H. Exclusão e Inclusão Social nas Sociedades Modernas: um olhar sobre a situação em Portugal e na União Europeia. Serviço Social & Sociedade, n. 106, p. 219-240, 2011.

5. ALVINO-BORBA, A., MATA-LIMA, A., MATA-LIMA, H. Desafios Ambientais e Estratégias para Desenvolvimento de Investigação e Programas de Intervenção Social. Ambiente & Sociedade, v. 15, n. 1, p. 147-155, 2012.

6. CHINO, M., NAKAYAMA, H., NAGAI, H., TERADA, H., KATATA, G., YAMAZAWA, H. Preliminary Estimation of Release Amounts of ^{131}I and ^{137}Cs Accidentally Discharged from the Fukushima Daiichi Nuclear Power Plant into te Atmosphere. Journal of Nuclear Science and Technology, v. 48, n. 7, p. 1129-1134, 2011.

7. COSTANZA, R., FARLEY, J. Ecological Economics of Coastal Disasters: introduction to the special issue. Ecological Economics, v. 63, n. 3, p. 249-253, 2007.

8. CUARESMA, J. Natural Disasters and Human Capital Accumulation. The World Bank Economic Review, v. 24, p. 280-302, 2010.

9. DAI, F.C., LEE, C.F., DENG, J.H., THAM, L.G. The 1786 Earthquake-Triggered Landslide Dam and Subsequent Dam-Break Flood on the Dadu River, Southwestern China. Geomorphology, v. 65, p. 205-221, 2005.

10. DATAR, A, LIU, J, LINNEMAYR, A, STECHER, C. The impact of natural disasters on child health and investments in rural India. Social Science & Medicine, v. 76, p. 83-91, 2013.

11. DAVIS, C., KEILIS-BOROK, V., KOSSOBOKOV, V., SOLOVIEV, A. Advance prediction of March 11, 2011 Great East Japan Earthquake: a missed opportunity for disaster preparedness. International Journal of Disaster Risk Reduction, v. 1, n. 1, p. 17-32, 2012.

12. EBEKE, C., COMBES, J.L. Do Remittances Dampen the Effect of Natural Disasters on Output Growth Volatility in Developing Countries? Applied Economics, v. 45, n. 16, p. 2241-2254, 2013.

13. FREITAS, C.M., CARVALHO, M.L., XIMENES, E.F., ARRAES, E.F., GOMES, J.O. Vulnerabilidade Socioambiental, Redução de Riscos de Desastres e Construção da Resiliência - lições do terremoto no Haiti e das chuvas fortes na Região Serrana, Brasil. Ciência & Saúde Coletiva, v. 17, n. 6, p. 1577-1586, 2012.

14. FRERKS, G., WARNER, J., WEIJS, B. The Politics of Vulnerability and Resilience. Ambiente & Sociedade, v. 14, n. 2, p. 105-122, 2011.

15. GIBBS, M.T. Resilience: what is it and what does it mean for marine policymakers? Marine Policy, v. 33, p. 322-331, 2009.

16. GUHA-SAPIR, D., VOS, F., BELOW, R., PONSERRE, S. Annual Disaster Statistical Review 2011: the numbers and trends. CRED, Brussels, 2012. Disponível em: <http://www.cred.be/sites/default/files/ADSR_2011.pdf>

17. HAVEN-TANG, C., JONES, E. Local Leadership for Rural Tourism Development: a case study of Adventa, Monmouthshire, UK. Tourism Management Perspectives, v. 4, p. 28-35, 2012.

18. HUAN, TC, BEAMAN, J, SHELBY, L. No-escape natural disaster: mitigating impacts on tourism. Annals of Tourism Research, v. 31, n. 2, p. 255-273, 2004.

19. IFRC. World Disasters Report 2003. International Federation of Red Cross and Red Crescent Societies (IFRC), Geneva, 2003.

20. IFRC. World Disasters Report 2010. Focus on Urban Risk. International Federation of Red Cross and Red Crescent Societies (IFRC), Geneva, 2010. Disponível em: <http://www.ifrc.org/Global/Publications/disasters/WDR/wdr2010/WDR2010-full.pdf>

21. INGRAM, J., FRANCO, G., RIO, C.R., KHAZAI, B. Post-disaster recovery dilemmas: challenges in balancing short-term and long-term needs for vulnerability reduction. Environmental Science & Policy, v. 9, p. 607-613, 2006.

22. IPCC. Climate Change 2007: Impacts, Adaptation and Vulnerability. Contribution of Working Group II to the Fourth Assessment Report of the Intergovernmental Panel on Climate Change (IPCC). Cambridge: Cambridge University Press, 2007.

23. ISO 14001. Sistema de Gestão Ambiental - Requisitos e linhas de orientação para a sua utilização (ISO 14001: 2004). International Standard Organisation (ISO), 2004. Disponível em: <http://www.anet.pt/downloads/legislacao/NP%20EN%20ISO%2014001%202004.pdf>

24. JACOBI, P.R., MONTEIRO, F. Social Capital and Institutional Performance: methodological and theoretical discussion on the water basin committees in metropolitan Sao Paulo - Brazil. Ambiente & Sociedade, v. 9, n. 2, p. 25-45, 2006.

25. JONES, J.A., SWANSON, F.J., WEMPLE, B.C., SNYDER, K.U. Effects of Roads on Hydrology, Geomorphology, and Disturbance Patches in Stream Networks. Conservation Biology, v. 14, n. 1, p. 76-85, 2000.

26. KAHN, M. The Death Toll from Natural Disasters: the role of income, geography, and institutions. The Review of Economics and Statistics, v. 87, n. 2, p. 271-284, 2005.

27. KLEIN, R.J.T., NICHOLLS, R.J., THOMALLA, F. Resilience to Natural Hazards: how useful is this concept? Global Environmental Change Parte B: Environmental Hazards, v. 5, n. 1-2, 2003.

28. KORTENHAUS, A., KAISER, G. Lessons learned from flood risk analyses at the North Sea Coast. Journal of Coastal Research, Special Issue 56, p. 822-826, 2009.

29. LIN, A., IKUTA, R., RAO, G. Tsunami Run-up Associated with Co-seismic Thrust Slip Produced by the 2011 Mw 9.0 Off Pacific Coast of Tohoku Earthquake, Japan. Earth and Planetary Science Letters, v. 337-338, p. 121-132, 2012.

30. LOAYZA, N., OLABERRÍA, E., RIGOLINI, J., CHRISTIAENSEN, J.R. Natural Disasters and Growth: going beyond the averages. World Development, v. 40, n. 7, p. 1317-1336, 2012.

31. MARAZZA D., BANDINI V., CONTIN A. Ranking Environmental Aspects in Environmental Management Systems: a new method tested on local authorities. Environment International, v. 36, p. 168-179, 2010.

32. MATA-LIMA, H. Human-Environment-Society Interactions: dam projects as a case example. Environmental Quality Management, v. 15, n. 3, p. 71-76, 2009.

33. MAUERHOFER, V. 3-D Sustainability: an approach for priority setting in situation of conflicting interests towards a sustainable development. Ecological Economics, v. 64, p. 496-506, 2008.

34. McENTIRE, D.A. Triggering Agents, Vulnerabilities and Disaster Reduction: towards a holistic paradigm. Disaster Prevention and Management, v. 10, n. 3, p. 189-196, 2001.

35. NAUDE, W. The Determinants of Migration from Sub-Saharan African Countries. Journal of African Economies, v. 19, p. 330-56, 2010.

36. NEDEL, A., SAUSEN, T.M., SAITO, S.M. Zoneamento dos Desastres Naturais Ocorridos no Estado do Rio Grande do Sul no Período 1989-2009: granizo e vendaval. Revista Brasileira de Meteorologia, v. 27, n. 2, p. 119-126, 2012.

37. NOY, I. The Macroeconomic Consequences of Disasters. Journal of Development Economics, v. 88, p. 221-231, 2009.

38. NOY, I & NUALSRI, T.B. The Economics of Natural Disasters in a Developing Country: the case of Vietnam. Journal of Asian Economics, v. 21, p. 345-354, 2010.

39. O'BRIEN, G., O'KEEFE, P., ROSE, J., WISNER, B. Climate Change and Disaster Management. Disaster, v. 30, n. 1, pp. 64-80, 2006.

40. OXLEY, MC. A "people-centred principles-based" post-Hyogo framework to strengthen the resilience of nations and communities. International Journal of Disaster Risk Reduction, v. 4, p. 1-9, 2013.

41. PARK, D.B., LEE, K.W., CHOI, H.S., YOON, Y. Factors Influencing Social Capital in Rural Tourism Communities in South Korea. Tourism Management, v. 33, p. 1511-1520, 2012.

42. POJASEK, R. Risk Management 101. Environmental Quality Management. v.13, n.3, p. 95-101, 2008.

43. REBELO, F. Um Novo Olhar Sobre os Riscos? O Exemplo das Cheias Rápidas (Flash Floods) em Domínio Mediterrâneo. Territorium, v. 15, p. 7-14, 2008.

44. SCHUMACHER, I, STROBL, E. Economic development and losses due to natural disasters: the role of hazard exposure. Ecological Economics, v. 72, p. 97-105, 2011.

45. SKOGDALEN, J.E., VINNEM, J.E. Quantitative Risk Analysis of Oil and Gas Drilling, Using Deepwater Horizon as Case Study. Reliability Engineering and System Safety, v. 100, p. 58-66, 2012.

46. SRINIVAS, H., NAKAGAWA, Y. Environmental Implications for Disaster Preparedness: lessons learnt from the Indian Ocean Tsunami. Journal of Environmental Management, v. 89, p. 4-13, 2008.

47. SU, B. Rural Tourism in China. Tourism Management, v. 32, p. 1438-1441, 2011.

48. TAKAHASHI, T., GOTO, M., YOSHIDA, H., SUMINO, H., MATSUI, H. Infectious Diseases After the 2011 Great East Japan

Esarthquake. Journal of Experimental & Clinical Medicine, v. 4, n. 1, p. 20-23, 2012.

49. TOMINAGA, L. K., SANTORO, J., AMARAL, R. Desastres Naturais: conhecer para prevenir. Instituto Geológico, São Paulo, 2009.

50. TOYA, H., SKIDMORE, M. Economic development and the impacts of natural disasters. Economics Letters, v. 94, p. 20-25, 2007.

51. TOPUZ, E., TALINLI, I., AYDIN, E. Integration of environmental and human health risk assessment for industries using hazardous materials: A quantitative multi criteria approach for environmental decision makers.Environment International, v.37, n.2, p. 393-403, 2011.

52. UN. Progress in the Implementation of the Programme of Action for Sustainable Development of Small Island Developing States: climate change and sea level rise. UN-Economic and Social Council, Report for the Secretary General, 19 - 30 April, 1999.

53. UNISDR. World Conference on Disaster Reduction. The United Nations Office for Disaster Risk Reduction (UNISDR). 18-22 Janeiro 2005, Kobe, Hyogo, Japan, 2005. Disponível em: <http://www.unisdr.org/2005/wcdr/wcdr-index.htm>

54. UNIVERSITY OF STRATHCLYDE. Environmental Assessment Method, 2000. Disponível em: <http://www.esru.strath.ac.uk/EandE/Web_sites/99-00/bio_fuel_cells/groupproject/library/environmentassess/text.htm>. Acesso: 08-08-2012.

55. VINK, G., ALLEN, R.M., CHAPIN, J., CROOKS, M., FRALEY, W., KRANTZ, J., LAVIGNE, A.M., LECUYER, A., MACCOLL, E.K., MORGAN, W.J., RIES, B., ROBINSON, E., RODRIQUEZ, K., SMITH, M., SPONBERG, K. (1998). Why the United States Is Becoming More Vulnerable to Natural Disasters. Eos, Transactions, American Geophysical Union, v. 79, n. 44, p. 533-537.

56. YASUHARA, K., KOMINE, H., MURAKAMI, S., CHEN, G., MITANI, Y., DUC, D.M. Effects of Climate Change on Geo-disasters in Coastal Zones and Their Adaptation. Geotextiles and Geomenbranes, v. 30, p. 24-34, 2012.

57. YODMANI, S. "Disaster Risk Management and Vulnerability Reduction: protecting the poor". Paper presented at the Asian

and Pacific Forum on Poverty, 5 - 9 February 2001, Manila, Philippines, 2001.

58.   WETZEL, R. Limnologia. Lisboa: Fundação Calouste Gulbenkian, 1996.

59.   WORLD BANK & UNITED NATIONS. Natural hazards, unnatural disasters: the economics of effective prevention. Washington, DC: The International Bank for Reconstruction and Development/The World Bank, 2010.

Chapter 4

The Vulnerability of the Elderly in Disasters: The Need for an Effective Resilience Policy

Airton Bodstein[I]; Valéria Vanda Azevedo de Lima[II];
and Angela Maria Abreu de Barros[III]

[I]Universidade Federal Fluminense
[II]Universidade Federal Fluminense
[III]Universidade Federal Fluminense

ABSTRACT

This article presents an analysis of the vulnerability of the elderly in un/natural disasters, since the elderly may have a higher degree of fragility in emergency situations among the vulnerable group. The aim is to support increasing the resilience of the elderly during critical events as part of Civil Protection and Defence. The data collection undertaken under the methodology of this research comprises senior citizens' rights as stipulated by Brazilian law, reports produced by official agencies, international documentation and general literature concerning this issue. The various Brazilian governmental projects and programmes focusing on the vulnerability of the elderly have not been effective as regards natural disasters. Therefore the results encourage Civil Defence to develop specific protocols on prevention, preparedness and rescue for the elderly.

INTRODUCTION

Longevity has been identified as one the most significant demographic changes of the 21st century. According to the 2012 report of the United Nations Population Fund (UNPF), in 2000 there were already more people worldwide aged 60 years or over than children under 5. Moreover, it is forecast that for the first time in the history of humanity, in 2050 there will be more people in this age group than children under 15.

There are currently 810 million people aged 60 or over worldwide, which equates to 11.5% of the global population, and the expectation is that this number will reach a billion in less than 10 years and more than double by 2050, affecting two billion people or 22% of the world's population. Japan is the only country in the world where more than 30% of the population is aged 60 or over, but, according to predictions, there will be 64 countries in 2050 in which the elderly population will make up 30% of the population.

Brazil, like other developing countries, is seeing a proportional drop in the population of young people and an increase in the proportion and absolute number of the elderly. The 2010 Demographic Census carried out by IBGE (Brazilian Institute of Geography and Statistics) highlighted a significant increase in the latter group over the last 50 years. In 1960, 3.3 million Brazilians were 60 or over and made up 4.7% of the population; in 2000, 14.5 million or 8.5% of Brazilians fell into this group. Over the last decade there has been a big jump: in 2010, this group represented 10.8% of the population, equating to 20.6 million people. In half a century (1960-2010), Brazilian life expectancy has increased by 25.4 years, rising from 48.0 to 73.4. Furthermore, over this period the average number of children per woman fell from

6.3 to 1.9, which is below the level for population replacement. It is forecast that in the period from 1950 to 2025 the elderly in Brazil will increase fifteen-fold, compared with a five-fold increase in the total population. If this occurs, the country will lie in sixth place as regards its cohort of the elderly, with roughly 32 million people aged 60 or over in 2025.

These changes, which have altered the Brazilian age pyramid, with a narrower base and a broader top, reflect the structure of an ageing population, which is a characteristic of more developed countries. However, ageing populations pan out differently in developing countries, where longevity is a recent phenomenon. As the World Health Organization states in its 2011 report *Global Health and Aging* (2011), it took more than 100 years in France for the percentage of the elderly aged 65 to rise from 7% to 14%, whereas in societies like Brazil, China and Thailand, this selfsame demographic path will be completed in around two decades. The prediction is that in 2050, 80% of the elderly will live in low income countries. At present, for each three people aged 60 or over, two of them live in developing countries, and the outlook for 2050 is that this number will rise to almost four out of five people. For 2020, the expectation is that the total number of people aged 60 or more will reach a billion, of whom 710 million will live in developing countries. What this means is that ageing populations in countries like Brazil represent a real challenge, both for politicians and society at large, especially for families who are legally responsible for caring for the elderly.

In Brazil, the first specific law to ensure the rights of the elderly was the National Senior Citizens Policy (*Política Nacional do Idoso*, PNI) (Law No 8.842 of 4th January, 1994), which was later regulated by the Senior Citizens' Statute (*Estatuto do Idoso*) (Law 10.741/2003), which states: "Families, society and the State have a duty to guarantee senior citizens all citizenship rights, ensuring they take part in the community and defending their dignity, well-being and right to life" (art.3, indent I). In turn, the "active ageing" policy championed by the World Health Organisation is based on the premise that ageing well is part of a collective construction which should be facilitated by public policies and access to healthcare throughout people's lives. This means "optimizing opportunities for health, participation and security in order to enhance quality of life as people grow older" (WHO, 2005).

Nevertheless, the targets laid down in laws and policies aimed at the elderly are still a long way from being met, chiefly in developing countries, where the shortage of material, human and technological resources, amongst others, is rendering the problems that affect the elderly population difficult to address whilst overburdening national infrastructures.

Faced with this reality, this work endeavours to analyse the vulnerability of the elderly in the context of disasters as part of Civil Protection and Defence, in the belief that within the group of vulnerable people – "children, pregnant women, the elderly and the disabled", as defined in Law 12.608, of April 2012, and in the Civil Defence Glossary (1998) –, the elderly are particularly at risk in emergencies. This is because in addition to the decline in functional capabilities stemming from the ageing process, multiple factors contribute to reducing senior citizens' resilience, such as illness, obesity, disability and living in at-risk areas, amongst others. Additionally, accidents at home or outside the home, together with illness and functional limitations, make the elderly even more vulnerable in disasters, as variables such as risk perception, state of alertness, attention, agility and mobility are compromised and hinder responses in these situations.

We have tried to systematise existing knowledge of the area, with the overall objective of contributing towards increasing the resilience of the elderly population as part of Civil Protection and Defence. To do so, we have researched how the phenomenon of population ageing raises new challenges in terms of protection measures. We have identified the causes for the vulnerability of the elderly and how they give rise to limiting factors when addressing disasters. The studies have been limited to natural disasters, given their increased intensity in recent years and the serious consequences they engender.

VARIABLES THAT IMPACT ON THE INCREASED VULNERABILITY OF THE ELDERLY

Functional Decline

A decline in functional ability as a whole can be seen during the natural and progressive ageing process. Variables such as risk perception, state of alertness, attention and mobility reduce gradually, considerably increasing the individual's vulnerability and the likelihood of becoming a victim of a critical event. Different factors and processes – physical, economic, social, psychological and physiological – all contribute to this vulnerability. Functional decline has been identified as the main manifestation of senior citizens' vulnerability, as shown by a state of fragility encompassing functional ability, balance and mobility, cognitive functioning, sensory deficiencies, emotional conditions/ presence of symptoms of depression, availability and suitability of family and social support, environmental conditions and nutritional state and risk (LACAS; ROCKWOOD, 2012).

Bearing in mind that health and functional state are key factors in determining the individual's response capability, and that, according to the World Health Organisation (WHO, 1946), "health is a state of complete physical, mental and social well-being and not merely the absence of disease or infirmity", equivalence between wellbeing and functionality can be assumed. Both represent the presence of autonomy (individual decision-making ability and command over actions, establishing and following own rules) and independence (ability to complete something using own resources), enabling individuals to look after themselves and their lives (MORAES, 2012). It can also be inferred that as the ageing population comprises a rather heterogeneous age group, with ages ranging from 60 to more than 100, the elderly population features differing degrees of autonomy and independence. In this way, the impact, direct or indirect, of critical events can influence the lives and health of this age group in differing forms and to differing intensities.

Accidents

The functional decline of the elderly can be further exacerbated by the consequences of accidents, both in the home and in public spaces. This increases their vulnerability in disasters, as these consequences may compromise the required mobility and agility, making it difficult for both themselves and rescue teams to react when a disaster strikes. Moreover, the vulnerable condition of the elderly means that the consequences of accidents are much more serious for them than for the population from other age groups, as they tend to have to spend longer in hospital and rehabilitation, and tend to be at a higher risk of dependency and death afterwards. With this in mind, it is worth stressing the prevalence of falls, which can greatly reduce senior citizens' quality of life, contributing to a drop in functional capacity, autonomy and independence.

In Brazil, in accordance with data from the Health Ministry (BRAZIL, 2009), falls and their consequences for the elderly have reached epidemic proportions, culminating in unfavourable outcomes, such as fragility, death, institutionalisation and a general worsening of health. Some factors that predispose the elderly to falls are physical fragility, use of medication which can lead to alterations in balance and/or vision, and a host of conditions, such as osteoporosis, for instance. However, falls may also occur due to a lack of prevention, be it at home, in institutions or in the communities where the elderly live.

The gravity of this scenario means that preventing falls is seen as a challenge to population ageing. According to the WHO's Global Report on Falls Prevention in Older Age (2010), it is estimated that 28% to 35% of over 65-year-olds experience falls every year. For the over 70s, the estimate ranges between 32% and 42%. The study stresses that in addition to causing significant physical and psychological limitations, which compromise senior citizens' quality of life and health, the injuries arising from falls, particularly fractures, have a critical economic impact on families, the community and society. Fractures, which lead to a high post-surgery mortality rate, can be linked to a range of factors both connected to the patients themselves and to their environment, such as unsuitable pavements, dim lighting, carpets and furniture location. With this in mind, the WHO report (2010) considers that the main factors to protect against falls are related to behavioural changes geared toward a healthy lifestyle and environmental modifications, such as adapting homes and improving public roads, for example.

Ageing and Urbanization

The 2010 IBGE Demographic Census shows that in 78 Brazilian municipalities, the elderly population already represents 20% of the population. In other words, one in five people are 60 years of age or over. More than half (53.2%) of the roughly 20.6 million people over 60 live in the most populous municipalities with more than 100 thousand inhabitants.

Bearing in mind that ageing and urbanisation are two worldwide trends, society needs to rethink the place of the elderly in towns and cities, in terms of infrastructure and services that are able to meet the demands of this population. The growth in the number of senior citizens living in towns and cities becomes problematic when the urban area's infrastructure and services are deficient, and there is environmental pollution and an increase in irregular occupations, all risk factors that intensify the vulnerability of the elderly to natural threats. Public spaces, buildings, open spaces, transport systems and housing represent the main characteristics of a city's physical environment. These have a major influence on personal mobility, and should theoretically be conceived so as to minimize the incidence of falls and injuries, seek to provide security against crime and encourage positive behaviour in relation to health and social participation. A city that is tailored to the reality of ageing contributes toward safe mobility, social participation, maintaining functional capacity and autonomy, and to helping increase resilience when disasters strike.

In this way, in addition to concerns about the conditions of public spaces - such as adaptations to the public transport network, traffic signs, pollution and the provision of specific areas to facilitate autonomy, independence and access for senior citizens -, the urban question must encompass the architecture of buildings, domestic spaces and provision of leisure and sport facilities that are compatible with the needs and capabilities of senior citizens.

It is worth stressing that the topic of urbanisation is the subject of the National Civil Protection and Defence Policy (PNPDEC), Law 12.608, enacted in April 2012, which has brought about considerable advances in Civil Defence and the way in which this dovetails with Urban Planning.

"NATURAL" DISASTERS OR DISASTERS WITH NATURAL CAUSES

The term "natural disasters" is widely used in Brazil both by the media and government bodies directly responsible for risk management and disasters in Brazil, as stated in all the official documents of the National Secretariat of Civil Defence of the Ministry of National Integration, as well as by the equivalent bodies at state and federal level. It is well worth highlighting that this designation was and still is used in the former Codification of Disasters, Threats and Risks – CODAR, now the COBRADE - Brazilian Classification and Codification of Disasters – to classify the origins of disasters.

In contrast, the International Strategy for Disaster Reduction - UNISDR and specialists in the area insist that a disaster is characterised by a highly abnormal situation and therefore can never be considered as a natural event.

Despite this stance of UNISDR, which was formulated some years ago now, the Ministry of National Integration in its Normative Ruling No 1, of 24th August 2012, classifies disasters as technological or natural, depending on the origins or primary causes. Classified as natural disasters are those "that may imply human losses or other health impacts, damage to the environment, interruption of services, and social and economic disturbances". In this article, the terms "disasters of natural origin or natural causes" are used, as suggested by UNISDR.

The serious consequences of disasters with natural causes in recent years highlight the considerable shortcomings of safety systems, particularly in developing countries. The data shows that the majority of the 3.3 million deaths caused by disasters in the last 40 years have been in poor countries. These countries also suffer the consequences of disasters for longer, insofar as they do not possess the ability to recover swiftly, which further worsens conditions of vulnerability (UNISDR, 2012).

According to UNDP – United Nations Development Programme (2012), "in Brazil, the myth that the country is immune to natural disasters has lost ground. In January 2011, for example, intense rainfall in the Serrano region of Rio de Janeiro caused the worst landslide in the country's history". For Carlos Nobre, Secretary for Research and Development Policies and Programmes at the Ministry of Science, Technology and Innovation – MCTI, this episode was the wake-up call for Brazilian perceptions of major disasters: "It became obvious to managers and to the population that the prevention angle needs to be emphasised. This was the conjuncture that changed our perspective forever: prevention is fundamental" (CASTRO, 2012). This thinking corroborates a study published previously by Masato Kobiyama et al (2006), according to which the increased impact of disasters with natural causes is mainly due to the increase in population, unregulated occupation due to the intense process of urbanisation and industrialisation, with the lack of investment in the prevention phase at the root of the problems faced.

Valencio et al. (2009) Stress That,

Of the various possible interpretations on what are designated disasters, one has to be taken into account in Brazil in particular; that is, what is recognised as a disaster at the institutional civil defence level is first and foremost a phenomenon of public perception of a vulnerability in the State's relationship with society when faced with the consequences of a threat it has not managed to prevent or has not been able to minimise the damage sufficiently. (p.5)

According to Debarati Guha-Sapir, one of the top specialists in the world on disasters and director of the Centre for Research on the Epidemiology of Disasters (CRED), a body that provides the UN with annual data on worldwide victims and is the point of reference for this topic, "there is no political will in Brazil to prepare the country to deal with natural disasters". At a press conference at the UN to present new statistics on the number of victims from disasters with natural causes in the world, she was categorical: "Brazil has enough money to deal with the problem of natural disasters and could have set up a prevention system years ago. But the overwhelming reality is a lack of political will" (CHADE, 2009).

As regards disaster prevention technology, the data shows that Brazil has acted reactively and not preventively. In 1966, straight after the intense rainfall in Rio de Janeiro that year, the Geotechnics Institute was set up. In 1975, it became the Geotechnics Superintendency and, in 1988, it began monitoring hillsides in the Rio de Janeiro municipality. Next, the Geo-Rio Foundation was established in 1992, whose disaster alert system is now called the Rio Alert System. It is worth stressing that since this Foundation was created there has been a significant drop in the number of deaths caused by landslides in Rio de Janeiro city. Another example of reactive action is the fact that it was only after the major disasters that hit Bumba in Niterói (2010) and in the Serrana Region in Rio de Janeiro state (2011), which resulted in more than a thousand victims, that the National Centre for Monitoring and Natural Disasters Warning (CEMADEN/MCTI) was set up and the National Centre for Risk and Disaster Management (CENAD/MI) was restructured in July 2011.

As for the systematisation of data on disasters in Brazil, which became available in 2012 when the Brazilian Natural Disasters Atlas 1991-2010 was published, it is of note that amongst the research limitations identified by the document were the variations and inconsistencies in recording human, material and economic damage, which to an extent undermines the historic database guiding the National Civil Defence System. The absence of specialised professionals at municipal level and the consequent lack of agreement and standardisation of the information published in the documents which record disasters also contribute to the shortcomings of the compiled data.

The Atlas identifies twelve natural phenomena related to the relevant disasters nationwide that were recorded in the five Brazilian regions over the twenty years under consideration: drought; rapid flooding; gradual flooding, hurricanes and/or cyclones; tornados; hail; frost; fire; mass movements; river erosion, linear erosion; marine erosion. Of all the disasters recorded between 1991 and 2010, drought corresponded to 16,944 entries (54%); rapid flooding and water-logging was second in terms of the highest incidence, with a total 6,771 records (21%); gradual flooding corresponded to 3,673 records (12%); hurricanes and cyclones, and hail, came next with 2,249 and 1,369 entries respectively and equated to 7% and 4%. The other disasters with natural causes - linear, marine and river erosion, forest fires, mass movements, tornadoes and frost - were of little import over the timeframe under consideration. They were therefore classified in the category "Others", with 903 incidences, representing 2% of the total records.

In this way, there has been an increase in the incidence of disasters, according to the data presented: of a total 31,909 disasters, 8,671 (27%) took place in the 1990s and 23,238 (73%) in the 2000s. Taking into account the shortcomings in record-keeping, as a trend it can be said that disasters have growth potential. In its final considerations, the Atlas concludes that the historic record of disasters in Brazil exposes the vulnerability of the Brazilian population to extreme situations related to climate phenomena and observes that:

A risk culture needs to be created that still does not exist in Brazil so that citizens are prepared to participate in the decision-making process. This measure can be made viable by providing access to quality information and by the main social stakeholders exchanging thoughts and reflections as part of a drive for participation and involvement of all sectors of society (p.91).

In 2011, according to official data from the Brazilian Natural Disaster Yearbook, 795 disasters with natural causes were recorded, which caused 1,094 fatalities and affected 12,535,401 people. Although the Southern Region was most affected by disasters (6,855,449 affected people), the region that suffered the greatest impact as a result of the power of destruction was the South East. The number of fatalities witnessed in this region is 7.29 times higher than that in the four other areas, due chiefly to the event that hit the Serrana Region of Rio de Janeiro, which represented 87.95% of total fatalities. Of the total number of affected persons (12,535,401), flooding was the disaster that affected the Brazilian population the most (56.19%) and was also the one that caused the highest number of fatalities (47.35%).

Of note is the fact that the natural right to life and wellbeing has been formally recognised by the Constitution of the Federative Republic of Brazil. It is the responsibility of Civil Defence to guarantee this right, specifically in disaster scenarios. Civil Protection and Defence in Brazil is organised under a system called the National Civil Protection and Defence System (SINPDEC), comprising the bodies and institutions of the Federal administration, the States, the Federal District and the Municipalities, as well as public and private bodies of note in the civil protection and defence sphere. The National Secretariat of Civil Defence - SEDEC, part of the Ministry of National Integration, is the central body of this system, tasked with coordinating civil protection and defence actions throughout the country (Law No. 12.608, of 11th April 2012).

Reducing the occurrence and intensity of disasters, which is the general objective of Civil Defence, encompasses prevention, mitigation, preparation and recovery actions and is multisectoral in its approach at the three levels of government - federal, state and local -, and with high levels of community involvement. However, as Masato Kobiyama *et al* (2006, introduction) state,

IN BRAZIL, NATURAL DISASTERS ARE TREATED IN A SEGMENTED FASHION AMONG THE DIFFERENT SECTORS OF SOCIETY. IN RECENT YEARS, THE DAMAGE CAUSED BY THESE PHENOMENA HAS INTENSIFIED DUE TO POOR URBAN PLANNING.

For the authors, integrated actions between the community and universities are crucial in minimising the effects of natural disasters, and they posit that the knowledge produced in the academic community should be passed on to society and used in preventive projects in an organised manner. At the local level, they suggest setting up community groups empowered to act before, during and after the event, thus helping civil defence management bodies.

The Elderly in Disasters

Sources for consultation dealing specifically with the elderly in the context of disasters proved insufficient for a more complete approach. Nevertheless, based on the information to which we had access, it is apparent that due to their vulnerability the elderly remain one of the hardest hit groups in critical situations.

The issue of the vulnerability of the elderly in disasters is the topic of the report *Older people in disasters and humanitarian crises*:

guidelines for best practice, by HelpAge International – HAI (2000), a global network of not-for-profit organisations that has been working with humanitarian agencies for over 20 years to address the special needs of the elderly in development projects for emergencies. The report has suggested ways in which the capacities and contributions of this social group can be bolstered. It includes guidelines on helping to understand and address the special needs of senior citizens in these situations, based on a broad range of research carried out in Asia, Africa, Europe and the Americas and many years' experience of disasters. When disasters strike, according to the document, specific elderly-related protocols are needed which require guidelines for, among others: evacuating persons with limited mobility such as residents of nursing homes; emergency shelters with no physical barriers; access to medication in a timely manner; availability of carers to assist with daily chores and tasks; access to support equipment such as walking sticks, wheelchairs and walking frames; and access to medical equipment, such as oxygen canisters.

It should be pointed out that HAI is the only international organisation working specifically to address the needs of the elderly, to defend their rights and to recognise the abilities and the contribution of the elderly in humanitarian crises. The fact that the organisation highlights the problems most commonly identified by the elderly in these situations and the needs to be addressed to improve serving this group is illustrative of how this reality has become an issue on the global stage. In its report, the HAI considers that for those working in developing countries, population ageing remains one of the questions most overlooked, and that in crisis situations the stated objective of the majority of organisations is to provide humanitarian assistance to communities, where possible to the most vulnerable. Yet research has clearly shown what can happen if senior citizens are not seen as being more vulnerable, as often they are excluded from social and economic recovery support programmes. This group urgently needs to be included in humanitarian response actions, as less than 1% of these actions target the elderly and disabled.

We can observe that the problems of the elderly take second place when it comes to government priorities and few non-governmental organisations (NGOs) include the elderly in their target populations. The common oversight is that senior citizens are very difficult to empower, they are not receptive to new ideas and they are incapable of effectively participating in community and economic activities.

Another observation is that the lack of awareness and information on the contribution of the elderly and their circumstances, problems and needs, creates negative images of old age. While awareness of the problems of the elderly has improved, such images and preconceptions remain, exacerbating this age group's "invisibility" and position as a non-priority. This goes back to the fact that "old age", as a symbolic concept, is interpreted by societies in accordance with their different cultural, historic and economic contexts.

The question of the vulnerability of the elderly in disasters, which has been broached in a number of international publications, will now be addressed. The article entitled "Elderly suffer cognitively during evacuation caused by natural disasters" (ISAUDE, 2012) presents the results of research carried out at the University of Pennsylvania School of Nursing, where 17 patients in long-term care were monitored. The patients were aged 86 on average and had fallen prey to a serious summer storm. They were all evacuated and transferred to other premises with different health care professionals and physical environments. The researchers remarked that during a disaster, physiological changes associated with ageing and the presence of chronic conditions made senior citizens more susceptible to illness or injury, or even death.

Referring to the Kobe disaster in 1995, the article "Recovery and Reconstruction after the Great Hanshin-Awaji Earthquake in Japan" (Murata, 2006) states that in some affected areas senior citizens were in the majority and 44% of the fatalities were over 65.

The effects of Hurricane Katrina, which hit the USA in 2005, on the elderly population were recorded in the article *Decline in Health Among Older Adults Affected by Hurricane Katrina* (22/01/2009), published by the Johns Hopkins Bloomberg School of Public Health. The study, which was undertaken in New Orleans, illustrates that in the year after the disaster, in addition to the increase in mortality, the health of survivors 65 or over showed a considerable decline. According to Lynda Burton, the study's main author, "there was a significant increase in the prevalence of patients with cardiac diagnoses, cardiac failure and sleeping problems".

Also with reference to Hurricane Katrina, the report *Current Status of the Social Situation, Wellbeing, Participation in Development and Rights of Older Persons Worldwide* (UNITED NATIONS, 2011) records that of the 1330 people who died, the majority were elderly. In the state of Louisiana, 71 percent of those who lost their lives were older than 60 years of age. The document states that the Louisiana Department of Health recorded that approximately 70 senior citizens living in care homes died on this occasion and many were abandoned by their carers during the disaster.

The same report also emphasises that a tsunami in Indonesia in 2004 mostly killed the elderly and children. In Europe during the 2003 heat wave the majority of deaths were among the elderly population, while just in France 70% of deaths were people aged 75 and over. As for the aforementioned earthquake that hit Kobe in 1995, more than half the fatalities were senior citizens and this group accounted for 90% of the subsequent deaths.

It is worth considering another complicating factor for the elderly: secondary disasters. According to the Civil Defence, outbreaks of leptospirosis stand out as one of the most prevalent secondary disasters in Brazil. Mortality resulting from this illness tends to increase amongst elderly patients and it also makes them more vulnerable when facing new situations of risk.

It is worth noting a fact that occurred in the recent past in Japan after the disaster caused by a tsunami in the Fukushima plant in March 2011. At the time, a group of more than two hundred Japanese retirees calling themselves the "Skilled Veterans' Unit", comprising engineers and other professionals, all over 60 years of age, got together of their own free will and volunteered to replace younger people in the work of trying to control the leak at the plant. They alleged that as they were elderly and at the end of their lives, they would not have time to develop cancer. In an interview with the BBC, the architect of the idea, Yasuteru Yamada, a 72-year-old retired engineer, stated: "On average, I've probably got another 13 to 15 years of life remaining. Even if I were exposed to the radiation, cancer needs 20 to 30 years to develop; therefore, we senior citizens have fewer chances of contracting cancer". This behaviour, a reflex of cultural conditioning, reflects the sense of responsibility felt by Japanese senior citizens towards society, which they see as a concept which only functions as a whole rather than individually.

FINAL CONSIDERATIONS AND RECOMMENDATIONS

Sources for consultation dealing specifically with the situation of the elderly in the context of natural disasters in Brazil proved insufficient for a more detailed approach to the topic. Nevertheless, based on the information to which we had access, we have observed how public policy instruments only partially address the needs of the elderly, as far as increasing longevity and improving quality of life are concerned. In emergency situations, demands to reduce the vulnerability of this social group go unheeded, and preventive actions that could increase their resilience throughout the ageing process are overlooked. This picture becomes even more troubling given the increased frequency and intensity with which disasters with natural causes are occurring throughout the world.

In accordance with some of the data obtained in our research, a trend can be seen in the growth potential for disasters over the last two decades, despite the shortcomings in record keeping. This being the case, it is crucial that citizens are made aware of the idea of building a risk culture, chiefly as regards to the inclusion of the elderly.

In Brazil, using the sources we consulted as a basis, we were not able to identify specific protocols regarding prevention, preparation and rescue to help the elderly in disasters, despite what is stipulated in the National Civil Protection and Defence Policy, and in Law 12.608, of 2012, where article 2, indent IV states: "Suggest procedures to help children, adolescents, pregnant women, the elderly and the disabled in disasters, whilst observing the applicable legislation" (our bold characters).

It has been shown that the elderly become more vulnerable in disasters as they are more liable to chronic illness and incapacity, resulting both from the natural decline in their functional capacity and from greater exposure to accidents This confirms the importance of having protocols that specifically take into account the needs and abilities of this group in the area of civil protection and defence. Looking at the whole spectrum of actions, accident prevention is of great importance, mainly in terms of falls, which considerably increase the fragility of senior citizens and which have taken on the dimension of an epidemic in Brazil. This fact shows that accident prevention policies are clearly needed for the elderly population as a means of indirectly increasing their resilience in disasters. Preventing accidental injuries, understanding their causes, adopting pedestrian protection measures, implementing policies to prevent falls and fires in the home (aimed at minimising their incidence) and providing assistance on questions of safety are measures that should be adopted by society as a whole.

As regards disaster prevention technology, the data shows that Brazil has acted reactively and not preventively. It can therefore be inferred that disasters arise from the combination of threats, conditions of vulnerability and capability, or insufficient measures to reduce the negative consequences of risk.

The rapid and unplanned urbanisation process of recent decades has been identified as having created economic and social consequences and also as having compromised cities' infrastructure in terms of meeting society's basic needs. This has increased the vulnerability of one part of the elderly population, who, through no choice of their own, have to live in potentially flood and landslide-prone areas. Bearing in mind that ageing and urbanisation are two global trends, it is urgent that both governments and society change the cultural benchmarks regarding the elderly who live in urban areas, based on a new vision of ageing.

Undoubtedly, a paradigm shift is underway. The elderly (or old age) are being reinvented in the current political and socioeconomic context in modern societies. Yet these changes are not taking place at an adequate pace. Despite the importance of public policies and the entire legal apparatus in favour of the elderly, with the National Senior Citizens Policy (PNI) and the Senior Citizens' Statute of particular note, effective action is still insufficient. Programmes, projects and actions should be implemented over the coming decades to guarantee and promote the autonomy and independence of the elderly, thus strengthening their resilience in the face of disasters with natural causes.

Lastly, the elderly do not represent a homogenous group, given that an array of specific conditions - financial resources, cultural differences, access to education, to leisure, to basic sanitation and healthcare services, for example - impact on their quality of life and influence their individual ageing process. Therefore, due to the specificities that distinguish them from each other, not all senior citizens have equal or similar needs, and this aspect also needs to be taken into account by public policies, including those on protection in disasters.

Finally, what can be inferred is that there is no such thing as the "elderly" as a universal category, but "senior citizens", who should be seen in all their multiple guises and specificities, particularly when implementing an effective policy for reducing disaster risks for all segments of society.

The results of this study highlight shortcomings that could be addressed by means of specific measures for helping the elderly. Recommendations include:

- promote the drawing up of specific protocols on prevention, preparation and rescue, targeted at the elderly in disaster contexts, by civil defence stakeholders at the three levels of government;
- develop an information system with statistical data on elderly victims (survivors and fatalities) of disasters with natural and technological causes, which could underpin future research in this area;
- methodologically evaluate the level of effectiveness of the whole framework of protection measures for the elderly developed within public and private institutions;
- In conjunction with Brazilian industry, stimulate the manufacture of easy-to-use products specifically for the elderly.

REFERENCES

1.   BRASIL. Ministério da Saúde. Saúde do Idoso: quedas de idosos, 2009. Disponível em:<http://portal.saude.gov.br/portal/saude/visualizar_texto.cfm?idtxt=33674&janela=1.> Acesso em: 04 jun. 2012

2.   CASTRO, Fábio de. "Nova legislação dará base científica à prevenção de desastres naturais, dizem especialistas». Revista Pesquisa FAPESP, 2012. Disponível em: <http://agencia.fapesp.br/16000> Acesso em 15 set. 2012

3.   CEPED UFSC. Atlas brasileiro de desastres naturais 1991 a 2010: volume Brasil. Florianópolis: CEPED UFSC, 2012. Disponível em: <http://150.162.127.14:8080/atlas/Brasil%20Rev.pdf.> Acesso em 04 set. 2012

4.   CHADE, Jamil. «Especialista em desastres naturais da ONU
      critica o Brasil». Jornal O estado de São Paulo [on line],
      22/01/2009. Disponível em <http://www.estadao.com.br/
      noticias/   vidae.especialista-em-desastres-naturais-da-onu-
      critica-o-brasil,311350,0.htm.>. Acesso em: 13 ago.2012
5.   DECLINE IN HEALTH AMONG OLDER ADULTS AFFECTED BY
      HURRICANE KATRINA. 22/01/2009. Johns Hopkins Bloomberg
      School of Public Health. Disponível em: <http://www.jhsph.
      edu/news/news-releases/2009/burton-hurricane-katrina-health.
      html.> Acesso em: 05 set. 2012
6.   GRUPO DE APOSENTADOS DO JAPÃO QUER ENFRENTAR
      RADIAÇÃO EM FUKUSHIMA. BBC Brasil - Multimídia. 31
      maio 2011. Disponível em: <http://www.bbc.co.uk/portuguese/
      multimedia/2011/05/110531_fukushima_aposentados_video.
      shtml.> Acesso em: 30 jul. 2012
7.   IBGE (2010). Síntese de Indicadores Sociais. Uma Análise
      das Condições de Vida da População Brasileira. Disponível
      em: <http://www.ibge.gov.br/home/estatistica/populacao/
      condicaodevida/indicadoresminimos/sinteseindicsociais2010/
      SIS_2010.pdf>. Acesso em 12 maio 2012
8.   ISAUDE.NET. Idosos sofrem cognitivamente durante evacuação
      causada por desastres naturais. 23/10/2011. Disponível em:
      <http://www.isaude.net/pt-BR/noticia/22109/geral/idosos-
      sofrem-cognitivamente-durante-evacuacao-causada-por-
      desastres-naturais.> Acesso em 17 ago. 2012
9.   LACAS, A.; ROCKWOOD, K. Frailty in primary care: a review of
      its conceptualization and implications for practice. BMC Med.,
      Londres, v. 10, n. 4, 11 Jan. 2012. Disponível em: <http://www.
      biomedcentral.com/1741-7015/10/4.> Acesso em: 07 nov. 2012
10.   MASATO, Kobiyama et al. Prevenção de desastres naturais:
      conceitos básicos. Curitiba: Ed. Organic Trading, 2006. 109p.
      Disponível   em: <http://www.ceped.ufsc.br/sites/default/files/
      projetos/Livro_Prevencao_de_Desastres_Naturais.pdf.>   Acesso
      em: 02 set. 2012
11.   MINISTÉRIO DA INTEGRAÇÃO NACIONAL. Gabinete do
      ministro. Instrução normativa N° 1, de 24 de agosto de 2012
      <http://www.in.gov.br/imprensa/visualiza/index.jsp?jornal=1&p
      agina=30&data=30/08/2012.> Acesso em: 22/01/2013

12. MORAES, Edgar Nunes. Atenção à saúde do idoso: aspectos conceituais. Brasília: Organização Pan-Americana da Saúde, 2012. 98 p. Disponível em: <http://apsredes.org/site2012/wp-content/uploads/2012/05/Saude-do-Idoso-WEB1.pdf.> Acesso em: 12 ago. 2012

13. MURATA, Masahiko. Recuperação e reconstrução depois do Grande Terremoto de Hanshin-Awaji no Japão-. Revista@local. glob - Pensamento Global para o Desenvolvimento Local. Número 3 - Ano de 2006, página 10. (colaboração (ONU/EIRD e IRP). Disponível em <http://www.delnetitcilo.net/pt/publicacoes-all/revista-do-delnet/local.glob-3/revista3_pt>. Acesso em: 07 out. 2012.

14. NAÇÕES UNIDAS (UNISDR). Estratégia Internacional para Redução de Desastres: o desastre sob o enfoque de novas lentes: para cada efeito, uma causa / Brigitte Leoni, Tim Radford, Mark Schulman; tradução Sarah Marcela Chinchilla Cartagena. São Paulo: CARE Brasil, 2012. Tradução de: Disaster through a different lens: behind every effect, there is a cause. Disponível em: <http://www.ceped.ufsc.br/sites/default/files/projetos/ guia_para_cobertura_jornalistica_em_rrd.pdf.> Acesso em: 07 nov.2012.

15. NAÇÕES UNIDAS (UNFPA), e HelpAge International. Envelhecimento no Século XXI: Celebração e Desafio. Nova York; Londres. 2012. Disponível em: <http://www.unfpa.org/webdav/ site/global/shared/documents/publications/2012/Portuguese-Exec-Summary.pdf.> Acesso em: 24 set. 2012

16. ONU/WHO (World Health Organization). Constitution of the World Health Organization. Basic Documents. WHO. Genebra, 1946. Disponível em: <http://www.direitoshumanos.usp.br/ index.php/OMS-Organiza%C3%A7%C3%A3o-Mundial-da-Sa%C3%BAde/constituicao-da-organizacao-mundial-da-saude-omswho.html.> Acesso em: 08 ago. 1012

17. OMS/WHO (2005). Envelhecimento ativo: uma política de saúde / World Health Organization; tradução Suzana Gontijo. – Brasília: Organização Pan-Americana da Saúde, 2005.   [Links]

18. OLDER PEOPLE IN DISASTERS AND HUMANITARIAN CRISES: GUIDELINES FOR BEST PRACTICE. HelpAge International

- HAI (2000). Disponível em: <http://www.helpage.org/download/4c4c9487c176.> Acesso em 02 jul. 2012

19. PROGRAMA DAS NAÇÕES UNIDAS PARA O DESENVOLVIMENTO – PNUD (2012). «PNUD e governo estudam parceria para prevenção de desastres naturais». Disponível em: <http://www.pnud.org.br/Noticia.aspx?id=3658.> Acesso em 13 ago. 2012

20. RELATÓRIO GLOBAL DA OMS SOBRE PREVENÇÃO DE QUEDAS NA VELHICE Secretaria de Estado da Saúde. São Paulo, 2010. Disponível em: <http://bvsms.saude.gov.br/bvs/publicacoes/relatorio_prevencao_quedas_velhice.pdf.> Acesso em: 23 out. 2012

21. SECRETARIA DA SAÚDE DO ESTADO DE SÃO PAULO. Vigilância e prevenção de quedas em idosos . Marilia C. P. Louvison e Tereza Etsuko da Costa Rosa (Ed.). São Paulo: SES/SP, 2010. Disponível em: <http://www.saude.sp.gov.br/resources/ccd/ publicacoes/publicacoes-ccd-saude-e-populacao/35344001_site.pdf.   > Acesso em: 14 out. 2012

22. UNITED NATIONS (2011). Department of Economic and Social Affairs, Population Division – DESA (2011). «Current Status of The Social Situation, Wellbeing, Participation In Development and Rights Of Older Persons Worldwide». Disponível em: <http://www.un.org/esa/socdev/ageing/documents/publications/current-status-older-persons.pdf. > Acesso em: 05 maio 2012

23. VALENCIO et al. 2009. Sociologia dos desastres: construção, interfaces e perspectivas no Brasil. São Carlos: Rima, 2009. Edição eletrônica em PDF. Disponível em <http://www.ufscar.br/neped/pdfs/livros/livro-sociologia-dos-desastres-versao-eletronica.pdf. > Acesso em: 21 set. 2012

24. VALENCIO, N. Desastres "naturais" ou genocídio velado? Subsídios para um exame sociológico do caso brasileiro. In: XXVIII CONGRESSO INTERNACIONAL DA ALAS, 28., Recife. Anais... Recife: UFPE, 2011. Disponível em <http://www.ufscar.br/neped/pdfs/anais/ALAS_2011-N._Valencio.pdf>. Acesso em 03 out.2013

Disaster Risk Reduction Knowledge of Local People in Nepal

Gangalal Tuladhar[1], Ryuichi Yatabe[2], Ranjan Kumar Dahal[3], and Netra Prakash Bhandary[2]

[1]Himalaya Conservation Group, Kathmandu, Nepal

[2]Department of Civil and Environmental Engineering, Graduate School of Science and Engineering, Ehime University, Matsuyama 790-8577, Japan

[3]Department of Geology, Trichandra Campus, Tribhuvan University, Tribhuvan, Nepal

ABSTRACT

Background

Nepal is highly vulnerable to natural disasters. A high proportion of the national GDP is lost every year in landslides, floods, and many other forms of disasters. A high number of human casualties and loss of public and private property in Nepal due to natural disasters may be attributed to inadequate public awareness, lack of disaster preparedness, weak governance, lack of coordination among the concerned government agencies, inadequate financial resources, and inadequate technical knowledge for mitigating the natural disasters. In this context, quite a few awareness and training programs for disaster risk reduction (DRR) have already been initiated in Nepal and their impact assessments are also already documented. However, effectiveness of the various implemented DRR programs is not yet evaluated through an independent study.

Results

The work presented in this paper explores local people's knowledge on disaster risk reduction (DRR). Altogether, 124 local people from 18 to 74 years of age from randomly selected 19 districts of Nepal were interviewed focusing on various questions on disaster information, disaster knowledge, disaster readiness, disaster awareness, disaster adaptation, and disaster risk perception. The collected response data were statistically analyzed using histogram and independent sample t-tests to examine the DRR knowledge of people. An independent t-test analysis (Table 1) suggests that there is no statistically significant gender-based difference in disaster knowledge, disaster readiness, disaster awareness, and disaster risk perception of the surveyed people. Disaster adaptation capacity of the local people was evaluated and more than 60 percent of the respondents were determined to adapt state of disaster in the community.

Conclusions

Findings of this independent research confirmed that the DRR education initiatives implemented in Nepal are not enough. The questionnaire survey results have pointed out at a few deficiencies in disseminating DRR knowledge in Nepal. We hope these findings will encourage the line agencies working in DRR issues in Nepal to modify their programs targeted for the local communities.

BACKGROUND

Disaster risk is expressed in terms of potential loss of lives, deterioration of health status and livelihoods, and potential damage to assets and services due to impact of existing natural hazard. Disaster risk reduction (DRR) is a systematic approach to identifying, assessing, and reducing disaster risk, and it helps minimize the vulnerability of a society or community (Maxwell and Buchanan-Smith [1994]; Bendimerad F [2003]; Kameda [2007]; Onstada et al. [2012]). It also prevents or mitigates the adverse effects of natural disasters, facilitating a sustainable development process. The Second World Conference on Disaster Reduction was held in Kobe (Hyogo), Japan in January 2005, which adopted the Hyogo Framework for Action (HFA) 2005–2015: Building the Resilience of Nations and Communities to Disasters. It has provided a unique opportunity to promote strategic and systematic approach to reducing vulnerabilities and risks. HFA states that all countries must use knowledge, innovation, and education to build a culture of safety and resilience at all levels. Moreover, it suggests that disasters can be reduced substantially if people are well informed and motivated about measures they can take to reduce vulnerability.

Nepal in the Himalayan region is one of the most disaster prone countries in the world. Because of its predominantly steep mountainous terrain in the north and low lying plains in the south, drained by steep and high current rivers originating from the Himalaya, and dominated by strong monsoonal rains, the country is overwhelmed by various natural disasters. The common disasters include landslides, debris flows, floods, earthquakes, snow avalanches, glacial lake outburst floods (GLOF), hailstorms, thunderbolts, cold waves, hot waves, and fire.

Knowingly and unknowingly poverty drives people to go live in high risk marginal areas of mountains and river valleys, which makes them vulnerable to disasters. On the other hand, heavy disaster losses such as during earthquakes and tsunamis or landslides and flood unexpectedly create poverty among a large number of people by destroying their houses, productive lands, other personal assets, and livelihood (Yamin et al. [2005]; Takeuchi et al. [2011]). Hence, poverty is both cause and consequence of disasters in under-developed or developing countries. Disaster risk reduction is particularly essential for sustaining the achievements of all kinds of development goals since it provides a safety net for the hard-earned development gains of a developing country (Holloway [2003]; Birkmann and von Teichman [2010]; Walshe and Nunn [2012]). In Nepal, it is a great challenge to protect infrastructure and public and individual properties from frequent landslide, flood, and earthquake disasters. Each year hundreds of people are killed and a large amount of public and private properties are destroyed in landslide, flood, fire, and avalanche disasters. Each large-scale disaster potentially sets the country back several years in terms of the development efforts. When scarce resources such as time, energy, expertise, and funding are suddenly diverted in relief and recovery work, the overall development activities are delayed significantly.

The disaster statistics of Nepal always motivate and justify the urgent need of DRR works in Nepal. Therefore, Nepal has also adopted HFA and so far the Government of Nepal (GoN) has assigned the national mandate towards DRR and mainstreaming the DRR in its various development as well as education programs. In Nepal, the World Disaster Reduction Campaign for 2006-2007 was initiated and many programs such amendment in school curricula for disaster risk education, community based disaster management in village level, disaster mitigation plans in district level etc. have been implemented. Similarly, raising awareness within school communities is the well implemented program in the schools of Nepal. This awareness activity include training of teachers; organizing disaster quiz competitions among schools and local youth clubs; school contests on disaster risk reduction knowledge; campaigning for disaster safety in communities; and turning school students into catalysts and initiators in many more community based disaster awareness activities. Results and progress of few disaster risk reduction (DRR) initiatives taken in schools and communities of Nepal were well documented (ActionAid [2011a], [b]). Recently, Nepal has also started to include disaster risk reduction into secondary and higher education system and curricula.

This article explores the effectiveness of DRR works in the rural communities of Nepal, and examines disaster knowledge of people, disaster preparedness, disaster awareness, disaster adaptation, and disaster risk. It also evaluates the effectiveness of recent DRR programs implemented by various international nongovernmental organizations and national nongovernmental organization (INGOs and NGOs) in the rural communities of Nepal.

Disaster Risk and Disaster Risk Reduction Initiatives in Nepal

Natural disasters in Nepal cause a significant impact on the national GDP particularly due to infrastructural damage, destruction of public and private properties, and loss of life. The loss of life and property in particular may be attributed to lack of public awareness, inadequate disaster preparedness, weak governance practice, lack of coordination among the government agencies, inadequate financial resources, and a low level of technical knowhow as well as skill in mitigating natural disasters. In recent years, however, development planners in Nepal seem to have understood the intimate link between the disasters and development strategies. In average, per day at least two people die in Nepal due to natural disasters (MoHA Ministry of Home Affairs et al.[2008]). A record of loss of human lives in various types of disasters in Nepal in the last 25 years (1986- 2011) is shown in Figure 1 (MoHA Ministry of Home Affairs [2003]; DWIDP Department of Water Induced Disaster Prevention [2006]; MoHA Ministry of Home Affairs et al. [2009]). The data are evident how severely the country has suffered from the natural disasters in the last two and half decades (1986-2011). In landslides and floods, the human casualty reaches as high as 288 per year. An existing data record in South Asia shows that Nepal stands third in annual average human deaths per million living population after Sri Lanka and Bangladesh.

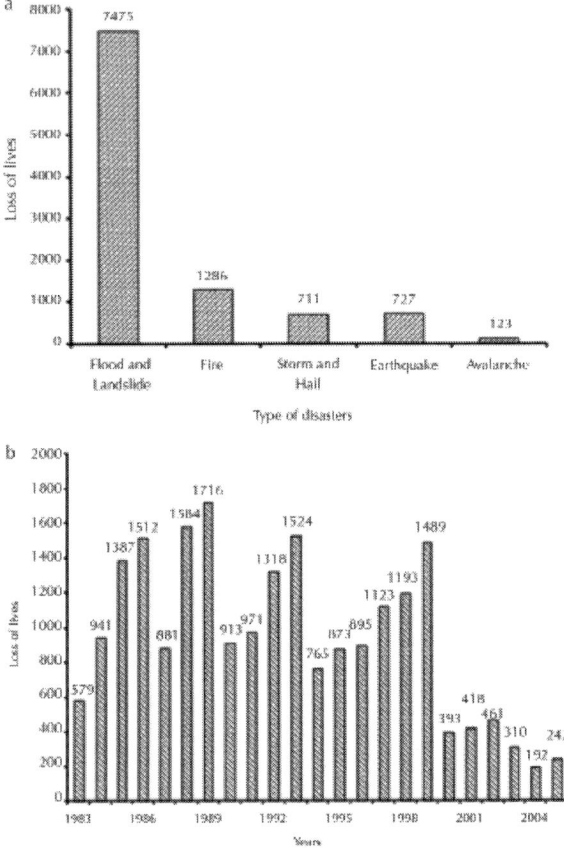

Figure 1: Loss of lives due to various disasters in Nepal between 1986 and 2005(a), and number of deaths due to disasters in Nepal from 1983 to 2005(b). Source: MoHA (2003), DWIDP (2006), MoHA et al. ([2009]).

These disaster statistics have always motivated and justified an urgent need of DRR works in Nepal. Therefore, Nepal is one of the 168 countries that have adopted the HFA. So far, the Government of Nepal (GoN) has assigned a national mandate towards disaster risk reduction and it's mainstreaming through various programs.

Following the HFA strategies, various international nongovernmental organizations working in the field of DRR have begun some ambitious programs designed to reduce people's vulnerability to natural disasters and build a stronger base of community-based disaster education. In Nepal too, especially after 2006 many programs have been introduced and implemented by various government and nongovernment agencies. A little change has also been made in the school-level curricula. Many disaster education-related programs have also been initiated (Figure 2) by both governmental and nongovernmental organizations (NGOs) (ActionAid [2011a], [2011b]; UNESCO United Nations Educational, Scientific and Cultural Organization and UNICEF United Nations Children's Fund [2012]; MercyCorp [2013]) in community levels.

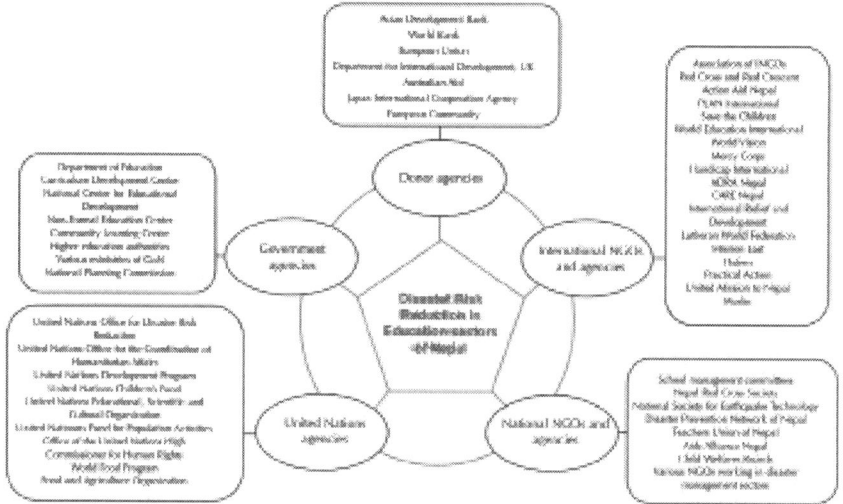

Figure 2: Framework for disaster risk reduction initiative in education sectors and implementation plan of Government of Nepal.

Raising DRR awareness level among the communities is one of the well-implemented programs in Nepal. The activities include teachers' trainings, disaster quiz competitions, youth club activities on DRR knowledge, disaster safety campaigns, and disaster drills. Establishing a sense of prevention in communities is another widely practiced DRR initiative in Nepal. For this, NGOs are involved in developing disaster education materials, coordinating for mainstreaming disaster risk reduction in national education system, and teaching youths, leaders, and parents the disaster risk reduction issues. Building earthquake safe communities and retrofitting existing structures are other areas of interest for the NGOs in Nepal towards building a disaster safe society. In this program, the government and NGOs are involved in assessing the vulnerability of school facilities, retrofitting school buildings, building earthquake-resistant schools, relocating schools in high disaster risk areas, and building new schools in low disaster risk areas. Results and progress of a few disaster risk reduction (DRR) initiatives taken in the schools of Nepal are well documented (Shiwaku et al.[2007]; ActionAid [2011a], [2011b]), and DRR has already been incorporated in the education system and school curricula (UNESCO United Nations Educational, Scientific and Cultural Organization and UNICEF United Nations Children's Fund [2012]).

METHODS

This study was intended to explore the level of DRR knowledge in local people and to examine the effect of DRR programs in Nepal on a number of aspects including risk perception, knowledge on available safety system in an event of disaster, preparedness of families and communities, and available disaster adaptation process up until now. The study also explores effectiveness of DRR implemented by various international nongovernmental organizations and national nongovernmental organization (INGOs and NGOs) in the rural communities of Nepal.

Data Collection

For this study, 19 districts of Nepal (out of 75) were randomly selected as sampling districts. During random selection, geographical distribution, development index and DRM activities of both government and nongovernment organization of each district were taken into consideration. The surveyed districts are shown in Figure 3. Also considered in the survey were activities of nongovernmental organizations in each district, disaster history (Aryal [2012]), rainfall-related disasters (Dahal and Hasegawa [2008]), and recent earthquake disaster (Dahal et al. [2012]). The study was conducted in assumptions that the local people are now gaining DRR knowledge through various trainings, awareness campaigns, and workshop programs organized by both national and international nongovernmental organization (ActionAid [2011a], [b], UNESCO United Nations Educational, Scientific and Cultural Organization and UNICEF United Nations Children's Fund[2012]).For the survey, a questionnaire sheet was prepared and a total of 124 local people (participants) from the randomly selected districts were asked to respond to the questions. The respondents consist of 15 percent female and 85 percent male with an age range of 18 to 74 years and mean of 38 years (SD = 11.8). Only 18 years (youths) or older from a variety of socioeconomic and cultural backgrounds were considered for the interview.

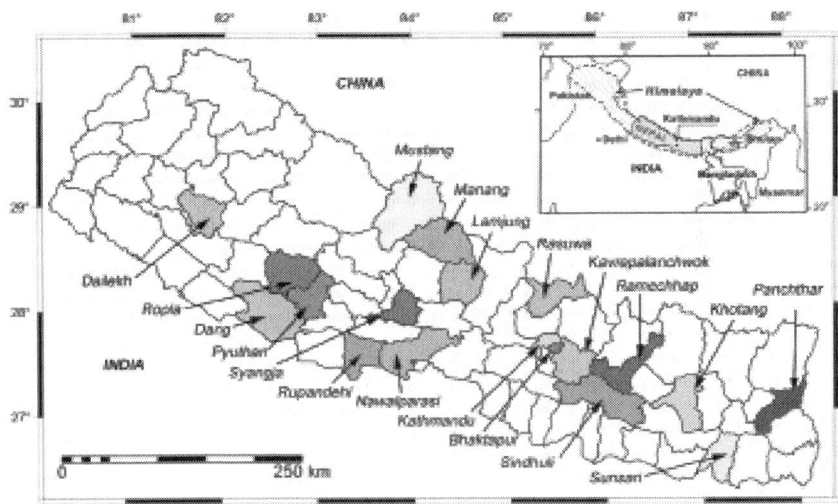

Figure 3: Location of 19 sample districts where randomly selected local people were interviewed.

The questionnaire survey criteria used in this study were adopted from the suggestions made in the available books and literatures (Kuroiwa [1993]; McMillan and Schumacher [1993]; Andrews et al. [1998]; Thorne [2000]; Henning et al. [2004]; Tanaka [2005]; Ronan et al. [2010]; Lekalakala[2011]), and they were embedded together within a single survey sheet.

Questions about various natural disasters were asked to assess the level of people's knowledge about these disasters. The participants' knowledge level was evaluated in terms of their understanding about the occurrence of floods, landslides, earthquakes, fires, high winds, hailstorm, drought, and extreme rainfall in five levels: (1) Never, (2) Rarely, (3) Sometimes, (4) Often, and (5) Always.

In addition, the survey participants were asked two sets of questions related to their feelings over the disaster and various issues of disaster management. . Major question "What are your feelings over the disaster?" was asked in the form of 9 statements. Similarly, 18 statements were asked to respond for another set of major question "What do you think about the following issues (18 statements) for disaster management". They were asked to indicate their responses in various statements (included in the two major question sets as most probable answers) in five levels: (1) Strongly disagree, (2) Disagree, (3) Agree, (4) Strongly agree, and (5) I do not know. Later in the analysis phase, the statements were categorized into five groups to explore knowledge of respondents on DRR as (i) Disaster-related knowledge, (ii) Disaster preparedness and readiness, (iiii) Disaster adaptation, (iv) Disaster awareness, (v) Disaster risk perception. A summary of the statements incorporated in the questionnaire survey is as follows.

Disaster-related knowledge
- I know when a disaster will occur
- I know disasters cannot be prevented
- I have participated in disaster risk education training or workshop

Disaster preparedness and readiness
- I think to come across a disaster and remain alive depends on our luck
- I know importance of disseminating experiences or knowledge of disaster
- I know government will provide enough facilities after disaster and we will not face any problem
- I am confident for reconstruction activities from government after disaster
- I know the importance of talking about disasters with neighbours, friends and colleagues
- I used to listen experts or DRR leaders who work or do activities for disaster management

Disaster adaptation
- I am aware of the shelter areas and open space in case of a disaster
- I have information about which government office needs to be contacted after the disaster

- I have knowledge about disaster prone area
- I am getting enough information from INGO/NGO about disaster adaptation
- I have knowledge about an evacuation area during a disaster
- I know the important of community activities for disasters risk reduction
- I know the life evacuation system in my locality

Disaster awareness

- I used to participate in voluntary activities for disaster awareness campaigns
- I am aware of retrofitting of buildings
- I used to prepare emergency bag for disasters
- I have a good relationship with my neighbours and community
- I think repair of road blockage and transportation break are important
- I give priority to disaster awareness in local, regional and national level
- I know recovery after disaster is a crucial work

Disaster risk perception

- I am very sure that large-scale disasters will certainly occur in next 10 years
- My locality is safe from all kinds of disasters
- I think my building is well designed and will withstand an earthquake event
- I am sure that my sleeping space is secure during and after disaster

Survey Procedure

Local representatives of the major political parties in Nepal, who have basic knowledge of disasters, were selected as enumerators, as they usually have a close acquaintance and a strong convincing relation with the local people. The enumerators were asked to select survey participants with basic education (that is, at least high school graduates) who could understand and answer the questions well. The survey was conducted more in presence of the enumerators themselves in an interview style for the clarity of the questions as well as answers of the respondents. In average, total time required for completing one survey was 20–30 minutes.

Method of Analysis

To examine overall DRR knowledge of local people, histogram analysis, bivariate correlations and independent sample t-tests was conducted. Basically, the descriptive analyses helped to examine the relationship between disaster risk reduction initiatives of government of Nepal and the local people's knowledge on DRR. Five key DRR issues were considered in our analysis: disaster knowledge, disaster readiness, disaster awareness, disaster adaptation, and disaster risk perception. Responses in these key issues were also evaluated with histogram analyses. A series of independent sample t-tests were also conducted to examine the effects of gender and disaster events. For this purpose, the five responses (Strongly disagree, Disagree, Agree, Strongly agree, and I do not know) were rephrased. For example, if a respondent responded strongly agree for all five DRR issues, it was considered that he/she well understood of the disaster knowledge, he/she was very ready to tackle the state of disaster, he/she is well aware for disaster risks, he/she can well adapt state of disaster and he/she is well perceived disaster risk. Similarly, if a respondent responded strongly disagree for all five DRR issues; it was considered that he/she has no idea of the disaster knowledge, disaster readiness, disaster awareness, disaster adaptation and disaster risk perception.

RESULTS

As mentioned in methodology, basically three kinds of analyses have been done to explore overall DRR knowledge of local people in Nepal. The effects of gender and disaster events were evaluated with independent sample t-tests and bivariate correlations. People's knowledge on DRR issues in Nepal was evaluated with histogram analyses. Disaster insecurity of local people was also evaluated from histogram plot. Results of analysis are given in the following headings.

Gender Effects on Disaster Risk Reduction Issues

Demographic factors always have some relationship with DRR process in a community. To explore this issue, preliminary analysis has been carried out on the basis of gender and age groups of the local people.

An independent t-test analysis (Table 1) suggests that there is no statistically significant gender-based difference in disaster knowledge, disaster readiness, disaster awareness, and disaster risk perception of the surveyed people, which can be understood from significance of t-test values greater than 0.05 (two-tailed) for almost all key disaster issues. Only for the case of awareness, the male participants were found more confused than the female, as indicated by less than 0.05 significance of t-test result.

Table 1: Statistical analysis of key disaster risk reduction issues

Key DRR Issues		Female		Male		t(124)	Sig.
Mean		SD	Mean	SD			
1	Knowledge: Well understood	42.33	10.50	33.67	5.51	1.27	0.27
	Knowledge: Understood	35.00	8.19	44.33	3.79	-1.79	0.15
	Knowledge: Not clear	16.00	5.00	11.67	4.62	1.10	0.33
	Knowledge: Confusing	5.33	5.51	7.00	0.00	-0.52	0.63
	Knowledge: No idea	1.67	2.89	4.33	1.53	-1.41	0.23
2	Readiness: Very ready	24.86	15.53	25.14	17.35	-0.03	0.97
	Readiness: Ready	39.14	15.74	32.57	16.49	0.76	0.46
	Readiness: Not ready	24.14	21.61	21.29	11.61	0.31	0.76
	Readiness: Confusing	7.71	4.42	12.29	10.29	-1.08	0.30
	Readiness: No idea	4.43	7.68	8.57	5.94	-1.13	0.28

3	Awareness: Well aware	21.71	13.56	21.71	11.76	0.00	1.00
	Awareness: Aware	42.00	10.50	42.43	5.80	-0.09	0.93
	Awareness: Not aware	22.71	10.34	16.14	5.90	1.46	0.17
	Awareness: Confusing	2.29	4.27	7.29	1.80	-2.85	0.01
	Awareness: No idea	11.43	4.93	13.14	7.52	-0.50	0.62
4	Adaptation: Well adapted	32.29	13.21	29.57	11.16	2.71	0.69
	Adaptation: Adapted	37.57	14.25	43.57	7.63	-6.00	0.35
	Adaptations: Not adapted	14.29	7.87	11.14	3.76	3.14	0.36
	Adaptation: Confusing	7.57	5.35	6.29	2.36	1.29	0.57
	Adaptation: No idea	8.43	7.44	10.00	4.55	-1.57	0.64
5	Perception: Well perceived	14.8	15.7	18.3	20.5	-0.582	0.582
	Perception: Perceived	9.0	11.6	14.3	8.5	-0.154	0.883
	Perception: Not perceived	44.8	20.3	28.8	18.0	1.18	0.283
	Perception: Confusing	22.5	17.0	24.0	9.5	-0.73	0.493
	Perception: No idea	9.3	15.3	15.5	15.0	-0.271	0.796

Tuladhar et al.

Tuladhar et al. Geoenvironmental Disasters 2015 2:5, doi:10.1186/s40677-014-0011-4

Likewise, when the people were asked about the use of media as a source of disaster information, it was found that the number of females using national television (that is, Nepal Television) is greater, but the males were found to prefer FM radios to learn about and get disaster information.

Disaster Risk Reduction Issues and People's Response

The DRR knowledge of local people was analyzed with people's response on five key DRR issues (disaster knowledge, disaster readiness, disaster awareness, disaster adaptation and disaster risk perception) considered in this research. Results for each issue are described in following sub-headings.

Disaster-Related Knowledge

Three main questions were asked to explore the level of disaster-related knowledge. More than 30 percent of the respondents were found to be familiar with the disaster-related facts (Figure 4). About 80 percent of them were found to agree with the importance of disaster risk-related trainings for them. This result indicates that the awareness campaigns of both governmental and nongovernmental organizations related to disaster knowledge in local level are in satisfactory level, and the people are rather positive about gaining disaster-related knowledge.

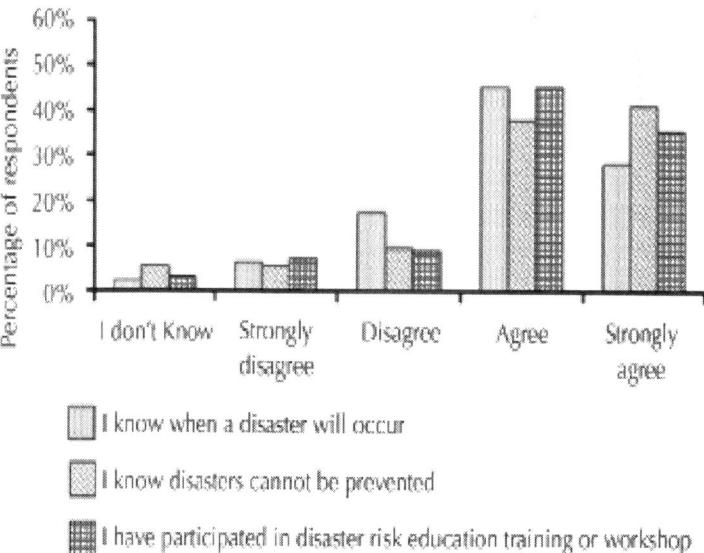

Figure 4: Evaluation of disaster-related knowledge in local people of Nepal.

Disaster Preparedness and Readiness Behaviour

Six main questions were asked to explore people's readiness behaviour towards the disasters. Out of these questions, there were positive responses for five questions and negative responses for two questions. More than 80 percent respondents do not think that the government has made enough preparations for DRR (Figure 5). They also do not agree that the government provides enough relief after a disaster. They also comment that there is a lack of governmental mechanism to support them after a disaster. About 25 percent respondents still believe that disaster and loss have direct link with their fate, while about 70 percent of the respondents are not convinced that governmental or nongovernmental institutions will initiate the post-disaster reconstruction activities. However, the respondents were found to be well motivated to talk about the disasters with their friends, colleagues, and neighbors. An overall impression about the readiness behavior of the people suggested that nearly 25 percent of the local people are still confused and are not ready to confront the disasters.

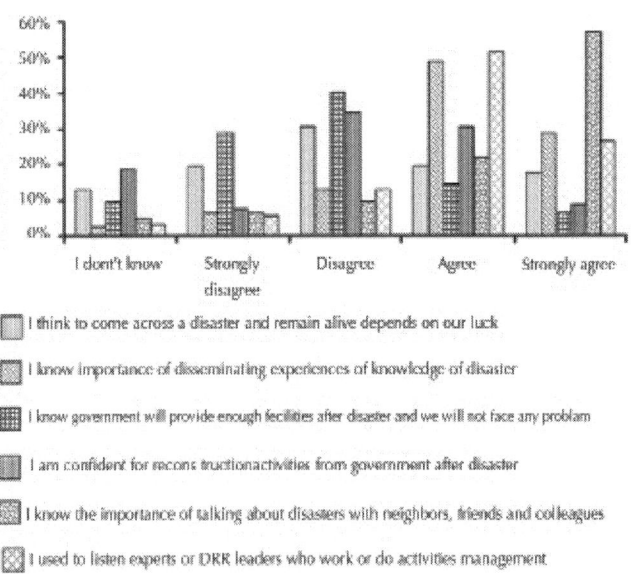

Figure 5: Readiness behaviour of people for disaster risk reduction.

Disaster AdaptationThe disaster adaptation capacity in the local people of Nepal was evaluated through seven main questions (Figure 6). In general, more than 60 percent of the respondents were determined to adapt state of disaster in the community. At present, although DRR programs and campaigns are being implemented and accomplished by various INGOs and NGOs, nearly 50 percent of the respondents was found negative on their activities, and respondents give little importance to the role of INGOs/NGOs in disaster information dissemination.

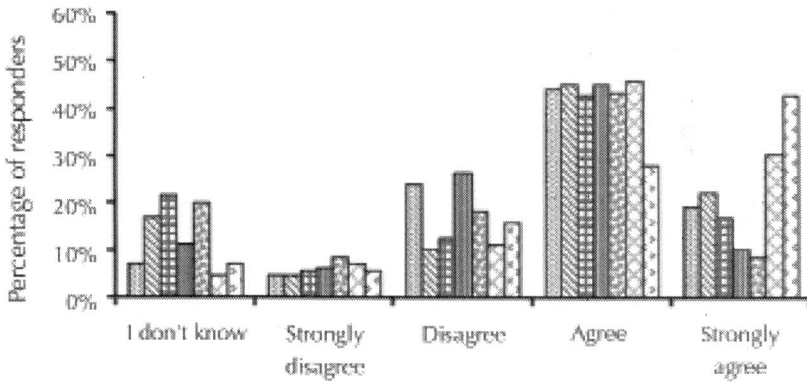

Aware of the shelter areas and open space in case of a disaster

Information about which government office needs to be contacted after the disaster

Knowledge about disaster prone area

Getting enough information from INGO/NGO about disaster adaptation

Knowledge about an evacuation area during a disaster

Community activities for disasters

Life in state of evacuation after the disaster

Figure 6: Response of the people to the various disaster adaptation systems in the community.

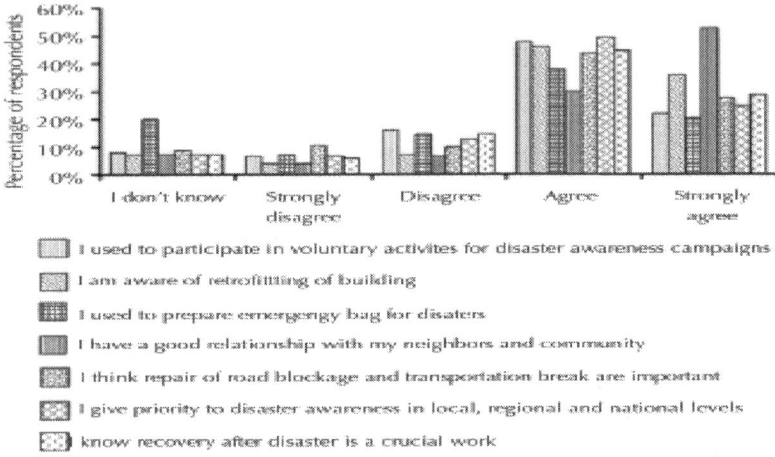

Figure 7: Response of the people to various disaster awareness related action in the community.

Disaster Awareness

Seven statements were asked to evaluate respondent's disaster awareness level (Figure 7). Only less than 20 percent of them were confused with the awareness activities for disaster risk in their community. This is a positive result for the governmental or nongovernmental institutions that are working for DRR issues in the community level. However, nearly 20 percent of respondents do not know or do not agree with the concept of disaster emergency bag. They emphasized that the concept of emergency bag is not practical for them.

Disaster Risk Perception

Four main questions were asked to the respondents so as to evaluate the risk perception. More than 75 percent of the respondents were found to be unaware of large-scale disasters in their communities (Figure 8) despite the fact that the annual disaster record of Nepal (see Figure 1) roughly indicates that major disasters occur in about every 10 years.

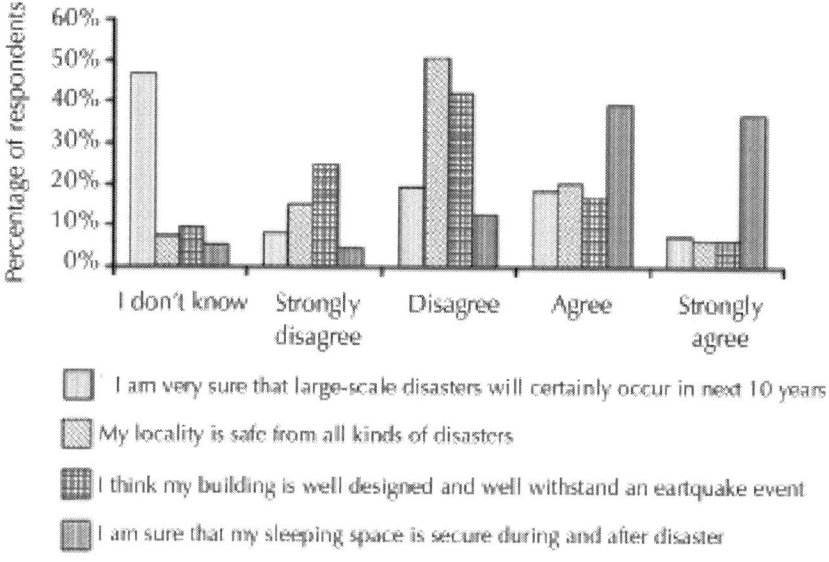

I am very sure that large-scale disasters will certainly occur in next 10 years

My locality is safe from all kinds of disasters

I think my building is well designed and well withstand an eartquake event

I am sure that my sleeping space is secure during and after disaster

Figure 8: Evaluation of risk perception of the people in the community.

Disaster Insecurity

People were asked about the level of insecurity they have from eight kinds of common natural disasters in Nepal. They responded in five levels of insecurity from the disasters. The responses clearly demonstrate their disaster risk perception. Most of the respondents feel that they are insecure from all kinds of disasters (Figure 9), but the maximum insecurity is associated with earthquake, storm, hail, drought, and extreme rainfall. Nearly 40 percent of the respondents feel that landslides may not be a problem for them, which in fact is a highly underestimated response. As most of the respondents are from mountainous areas, they must have a sound knowledge of landslide processes and associated disasters in their area. In case of floods also, the respondents were found to have a similar opinion. This clearly indicates that the DRR issues are either not being well protruded or are focused more on earthquake issues in the community level. Although many people are well aware disaster awareness programs, still one third of the respondents were worried for all kind of disasters and could not recognize major disaster problem in his/her area.

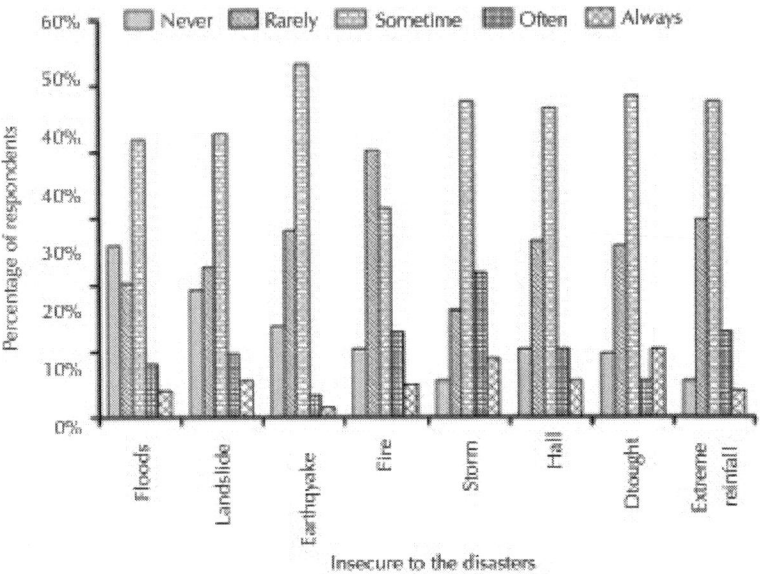

Figure 9: Level of anxiety from different kind of disasters in Nepal.

DISCUSSION

This study has helped to understand the status and importance of DRR knowledge dissemination process in rural communities of Nepal. Although the line agencies (that, governmental and nongovernmental institutions that have been involved in DRR activities in various communities of Nepal) claim that DRR concept and disaster education are now already functioning in the local communities and all local people have been gaining DRR knowledge through awareness campaigns, trainings, meetings, and so on, the ground reality indicates that the situation is still incoherent. In this work, people's knowledge in five key DRR issues was explored through a questionnaire survey on at least high school graduates, but the findings are not very encouraging. For example, one of findings indicates that many people are still obscured on awareness activities for disaster risk management in community. A few satisfactory results were obtained particularly concerning the status of people's knowledge in disaster. Although the level of knowledge of both male and female respondents in DRR issues is not different, many males were found to be still confused about procedures for raising disaster awareness. An analysis of the obtained results has shown that the local people lack accurate knowledge of disasters and their mitigation. Despite a fact that landslides and flood are most affecting and most frequent natural disasters in Nepal, many people were found to feel only less insecure from these disaster events. In Nepal, for the last 20 years, the information about earthquake disaster is well disseminated by various organizations (Dixit et al. [2013]), which might have resulted in positive consequences of people's increased awareness in earthquake-related disasters. Moreover, despite a fact that disaster education programs are doing good or satisfactory in Nepal (UNESCO United Nations Educational, Scientific and Cultural Organization and UNICEF United Nations Children's Fund [2012]), the survey results have shown that the people have a greater likelihood of feeling insecure about all kinds of disaster. Even today, most people do not have site-specific disaster knowledge, and the level of anxiety towards all kinds of disasters is the same. More than 30 percent of the respondents have answered that all of the eight kinds of disasters (flood, landslide, earthquake, fire, storm, hail, drought, and extreme rainfall) may occur sometime in their areas, which in fact is not a correct understanding of major disaster issues of his/her area.

Although the number of samples collected is not high, this study has pointed out a few deficiencies in the process of disseminating disaster risk reduction knowledge in Nepal. The findings in this work are expected to encourage the line agencies that have been working in DRR issues in the country towards modification in their programs intended for the local communities. We believe an independent research work, such the one done in this study, more clearly shows the overall status of people's knowledge as well as understanding in any relevant fields including DRR.

CONCLUSIONS

The questionnaire survey results obtained during the course of this field-based research work clearly point out at the deficiencies and limitations in the process of DRR knowledge dissemination to the local communities of Nepal. This work has attempted to evaluate specific outcomes in relation to disaster preparedness of the local people, their ability to identify and address the risk of disasters, and disaster adaptation practice. However, in many DRR issues, people are still not very clear and at the same time, they respond incorrectly. Likewise, she/he is well aware for disaster risks but she/he is not well familiar with the adapt state of disaster. This is a contradictory response obtained during survey. Through this study, it is understood that there are many DRR lessons the local people need to learn further, and that the existing DRR knowledge dissemination programs and processes need to be revised. Despite also having the Hyogo Framework for Action 2005–2015 (UNISDR United Nations International Strategy for Disaster Reduction [2004], [2007],[2011]) adopted, the survey results show that Nepal has not satisfactorily implemented the DRR education initiatives. The HFA well emphasizes the role of education in DRR, especially the need of disaster education for developing a culture of disaster prevention in communities and schools.

At the present political and social conditions, the disaster risk reduction programs run particularly by the INGOs or NGOs may sometimes also be misleading. As an important tributary of a broader sustainable development pathway, DRR must be integrated with the development process in the communities. It needs to be mixed up with the development process at different stages, and must become an integral part of the development activities. In DRR process, culture of safety can also be easily introduced if the communities are adequately educated, equipped, and resourceful through good governance. In reality, the existing DRR programs in Nepal are rich in jargons and they cannot be simply or easily implemented in communities. Through this study, we have clearly understood that people are not adequately aware even of simple disaster issues, and that in some cases, they are over reckoning too. In general, people need information that leads to action, for example, what they should do on their own today or tomorrow and who they should consult for any additional support. Another tragedy about DRR in Nepal is that its practice is badly overshadowed and blended into various hypothetical approaches; and to the worst, most of the resources are spent more in discussions, workshops, and hiring foreign consultants instead of taking immediate action in the field with whatsoever available indigenous knowledge. As a result, DRR mainstreaming programs are likely to fade out between the communities and the line agencies leading to a dilemma of DRR knowledge among the local communities.

One of the major challenges in DRR practice in a rural country like Nepal is implementation method, especially at personal level. The method of disaster education provides people knowledge and information and promotes the DRR measures. To achieve this, local people need to be encouraged to learn about disaster basics, readiness behavior, awareness programs, adaptation process, and risk identification techniques. To strengthen people's disaster risk awareness, proper and appropriate information should be disseminated to the local community leaders. Disaster education-related campaigns and programs may also provide a self-learning environment to the local people. School disaster education programs should also be encouraged in such a way that the community people can participate in the school disaster education program and the students can participate in community-based disaster awareness and adaptation activities. These activities may help increase the knowledge of DRR in the communities, which may lead the community and people to greater readiness for disaster risk reduction process.

AUTHOR'S CONTRIBUTION

GT collected the data and performed the statistical analysis. He also drafted the manuscript. RY, RKD and NPB give suggestions on methodology and developing questionnaire. All four authors read and approved the final manuscript.

ACKNOWLEDGMENTS

This research has been conducted with the extensive help from officials and staffs of Himalayan Conservation Group. Local political leaders are also acknowledged for their heartfelt support in data collection.

REFERENCES

1. ActionAid: *Disaster Risk Reduction through Schools: Learning from our experience 5 years on*. 2011.

2. ActionAid: *Disaster Risk Reduction through Schools: A Groundbreaking Project*. 2011.

3.   Andrews J, Benthien M, Tekula S (1998) Southern California Earthquake Center Outreach Report 1998: Public Awareness, Education and Knowledge Transfer Programs and Fiscal Year 1998 Activities. http://www.scec.org/research/98research/98andrews. pdf. Accessed on May 2012.

4.   Arya AS: Training and drills for the general public in emergency response to a major earthquake. [http://www.virtualref.com/ uncrd/1563.htm] *Training and Education for Improving Earthquake Disaster Management in Developing Countries* United Nations Centre for Regional Development, Nagoya, Japan; 1993, 103-114. http://www.virtualref.com/uncrd/1563.htm

5.   Aryal KR: The History of Disaster Incidents and Impacts in Nepal, 1900–2005. *Int J Disaster Risk Sci* 2012, 3(3):147-154.

6.   Bendimerad F (2003) Disaster Risk Reduction and Sustainable Development. World Bank Seminar on The Role of Local Governments in Reducing the Risk of Disasters, Held in Istanbul, Turkey, 28 April–2 May 2003, 57–75. http://info. worldbank.org/etools/docs/library/114715/istanbul03/docs/ istanbul03/05bendimerad3-n[1].pdf. Accessed on May 2013.

7.   Birkmann J, von Teichman K: Integrating disaster risk reduction and climate change adaptation: key challenges – scales, knowledge, and norms. *Sustainability Sci* 2010, 5(2):171-184.

8.   Dahal RK, Hasegawa S: Representative rainfall thresholds for landslides in the Nepal Himalaya. *Geomorphology* 2008, 100(3–4):429-443.

9.   Dahal RK, Bhandary NP, Yatabe R, Timilsina M, Hasegawa S: Earthquake-Induced landslide in the roadside slopes of east Nepal after recent September 18, 2011 earthquake. In *Earthquake-Induced Landslides*. Edited by Ugai K, Yagi H, Wakai A. Springer-Verlag, Berlin; 2012:149-157.

10.   Dixit AM, Yatabe R, Dahal RK, Bhandary NP: Initiatives for earthquake disaster risk management in the Kathmandu Valley. *Nat Hazards* 2013, 69(1):631-654.

11.   DWIDP (Department of Water Induced Disaster Prevention) (2006) Disaster Review 2006, http://www.dwidp.gov.np/uploads/ document/file/review_20120213035717.pdf.

12.   Henning E, Van Rensburg W, Smit B: *Finding Your Way in Qualitative Research*. Van Schaik Publishers, Pretoria; 2004.

13. Holloway A: Disaster risk reduction in Southern Africa. *Afr Security Review* 2003, 12(1):29-38.

14. Kameda H (2007) Networking disaster risk reduction technology and knowledge through Disaster Reduction Hyperbase (DRH). In: Proceedings of the Disaster Reduction Hyperbase (DRH) Contents Meeting, Kobe, Japan, 12-13 March. http://drh.edm. bosai.go.jp/Project/Phase2/1Documents/9_EXr.pdf.

15. Kuroiwa J: Peru's National Educational Program for Disaster Prevention and Mitigation (PNEPDPM). [http://www.virtualref. com/uncrd/1562.htm] *Training and Education for Improving Earthquake Disaster Management in Developing Countries* United Nations Centre for Regional Development, Nagoya, Japan; 1993, 95-102. http://www.virtualref.com/uncrd/1562.htm

16. Lekalakala MJ: Teachers perceptions about lesson planning to include a disaster risk reduction focus. In *Master's thesis, Disaster Management Training and Education Centre for Africa*. University of the Free State, South Africa; 2011.

17. McMillan JH, Schumacher S: *Research in Education*. A Conceptual Introduction, New York: Harper Collins; 1993.

18. MercyCorp (2013) Supporting the Role of Schools in Disaster Risk Reduction (SRSD). http://nepal.mercycorps.org/projects/ disaster-projects/srsd.php

19. Maxwell S, Buchanan-Smith M: Linking relief and development: introduction and overview. *IDS Bulletin* 1994, 25(4):1-19

20. MoHA (Ministry of Home Affairs): *Disaster Report of Nepal.* Ministry of Home Affairs, Government of Nepal; 2003.

21. MoHA (Ministry of Home Affairs), UNDP (United Nations Development Programme), EC (European Comission), NSET (National Society for Earthquake Technology-Nepal). (2008). National Strategy for Disaster Risk Management in Nepal. Kathmandu: MoHA, UNDP, EC, NSET. http://www. rccdm.net/index.php?option=com_docman&task=doc_ view&Itemid=215&gid=17

22. MoHA (Ministry of Home Affairs): Nepal Disaster Report In *The hardship and vulnerability. Ministry of Home Affairs, Government of Nepal and Disaster Preparedness Network-Nepal.* Jagadamba Press, Nepal; 2009.

23. Onstada PA, Danesb SM, Hardmanc AM, Olsonc PD, Marczakc MS, Heinsd RK, Croymanse SR, Coffeec KA: The road to recovery from a natural disaster: voices from the community. *Community Development* 2012, 3(5):566-580.

24. Ronan K, Crellin K, Johnston D: Correlates of hazards education for youth: a replication study. *Nat Hazards* 2010, 53(3):503-526.

25. Shiwaku K, Shaw R, Kandel RC, Shrestha SN, Dixit AM: Future perspective of school disaster education in Nepal. *Disaster Prev Manage* 2007, 16(4):576-587.

26. Tanaka K: The impact of disaster education on public preparation and mitigation for earthquakes: a cross-country comparison between Fukui, Japan and the San Francisco Bay Area, California, USA. *Applied Geography* 2005, 25(3):201-225.

27. Takeuchi Y, Mulyasari F, Shaw R: Chapter 4 Roles of Family and Community in Disaster Education. In *Community, Environment and Disaster Risk Management, vol 7*. Edited by Shaw R, Takeuchi KSY. Emerald Group Publishing Limited, UK; 2011:77-94.

28. Thorne SRN: Data analysis in qualitative research. *Evid Based Nurs* 2000, 3(3):68-70.

29. UNESCO (United Nations Educational, Scientific and Cultural Organization) and UNICEF (United Nations Children's Fund): *Disaster Risk Reduction in School Curricula: Case Studies from Thirty Countries*. France United Nations Educational, Scientific and Cultural Organization and United Nations Children's Fund, Peris; 2012.

30. UNISDR (United Nations International Strategy for Disaster Reduction) (2004) Hyogo Framework for Action 2005–2015. United Nations Inter-Agency Secretariat of the International Strategy for Disaster Reduction. Accessed on 25 November 2013 http://www.unisdr.org/2005/wcdr/intergover/official-doc/L-docs/Hyogo-framework-for-action-english.pdf. Accessed on 25 November 2013

31. UNISDR (United Nations International Strategy for Disaster Reduction): *Towards a Culture of Prevention: Disaster Risk Reduction Begins at School – Good Practices and Lessons Learned*. United Nations International Strategy for Disaster Reduction, Geneva; 2007.

32. UNISDR (United Nations International Strategy for Disaster Reduction: *Compilation of National Progress Reports on the Implementation of the Hyogo Framework for Action, HFA Priority 3, Core Indicator 3.2*. 2011.

33. Walshe RA, Nunn PD: Integration of Indigenous Knowledge and Disaster Risk Reduction: A Case Study from Baie Martelli, Pentecost Island, Vanuatu. *Int J Disaster Risk Sci* 2012, 3(4):185-194.

34. Yamin F, Rahman A, Huq S: Vulnerability, Adaptation and Climate Disasters: A Conceptual Overview. *IDS Bulletin* 2005, 36(4):1-14.

Trajectories of Social Vulnerability During the Soufrière Hills Volcanic Crisis

Anna Hicks[1] and Roger Few[2]

[1]School of Environmental Sciences, University of East Anglia, Norwich Research Park, Norwich NR4 7TJ, Norfolk, UK
[2]School of International Development, University of East Anglia, Norwich Research Park, Norwich NR4 7TJ, Norfolk, UK

ABSTRACT

When some active volcanoes enter into an eruptive phase, they generate a succession of hazard events manifested over a multi-year period of time. Under such conditions of prolonged risk, understanding what makes a population vulnerable to volcanic threats is a complex and nuanced process, and must be analysed within the wider context of physical events, decisions, actions and inactions which may have accentuated the social differentiation of impacts. Further, we must acknowledge the temporal component of vulnerability, therefore our analyses must go beyond a transitory view to an understanding of the dynamics of vulnerability, particularly how inherent socio-economic conditions drive vulnerability today, and how patterns of vulnerability shift during the course of a long-lived crisis.

INTRODUCTION

The complex, variable and dynamic nature of volcanic activity creates a multi-dimensional impact on people and assets, influenced by physical and social vulnerability and societal capacity to respond. Attempts to reduce volcanic risk require a detailed understanding of how these components interact to change risk and impact resilience. The challenges for society - as well as for analysis – become accentuated further in situations where volcanoes enter into a prolonged eruptive phase, when the ramifications for society and economy can take on a deeper, more sustained nature.

Through the 'Strengthening Resilience in Volcanic Areas' (STREVA) project, a series of 'forensic' studies of risk were carried out in Montserrat, an island that has experienced a long-lived volcanic crisis since 1995. This paper focuses on dimensions of vulnerability analysed through the forensic research. In doing so, it takes a view of vulnerability that not only examines the antecedent conditions that could preclude or catalyse disasters, but also analyses the changes in capacity of a population to recover and adapt.

Assessing Vulnerability in Volcanic Settings

In this study, we refer to 'vulnerability' as the potential to experience detrimental outcomes to wellbeing, lives and livelihoods, as a result of a hazard event - in this case a long-lived volcanic crisis. This usage of the term matches that within a body of critical social science work at the junction between political ecology, hazards research and development studies (e.g. Bankoff et al. [2004]; Wisner et al. [2004]; Cutter [1996]; Schipper and Pelling [2006]), which views vulnerability not just as a function of physical exposure to hazard but crucially also as a function of susceptibility to the effects of that exposure. Both components are inter-related and inherently 'social' in that it is social processes that largely determine different abilities to avoid, prepare for, withstand and recover from impacts of hazards (Wisner et al. [2004]). This approach to analysing vulnerability therefore requires attention to social structures (such as modes of governance and rules of land tenure, for example) as well as to patterns of variance in resources and livelihood assets at the individual and household level (Pelling [2003]; Few [2007]; Gaillard [2008]). Vulnerability is in this sense distinct from 'impact' (which is the actual effect of a hazard event), but the underlying idea we are proposing in this study is that in a post-eruption setting one can look at relative vulnerability as revealed through the prism of different impacts on different social groups.

Despite widespread recognition that assessments of vulnerability are essential to help design effective strategies for risk reduction to natural hazards, for volcanic risk, there remains disproportionate research focus on assessment of the hazard (Sword-Daniels [2011]). Studies directed towards vulnerability assessment have a tendency to be focused solely on physical vulnerability (i.e. the likelihood of physical exposure to the hazard) and, while this is an important component of volcanic risk analysis, it needs to be supported by further research to identify the differentiation and dynamics of societal vulnerability to volcanic hazards (e.g., Dibben and Chester [1999]; Wisner et al [2004]). Further, empirical research that *integrates* vulnerability data into volcanic risk assessments is virtually absent (as an exception, see Hicks et al. [2014]). This is likely a function of, a) the complexity of integrating qualitative and quantitative data sets; b) the challenges of effectively working in an interdisciplinary team to produce new knowledge, when disciplinary methodologies and epistemologies are seemingly incompatible, and c) a disciplinary mismatch of required time for data gathering (social scientific data, for example, usually requires a longer time to obtain than many forms of physical scientific data).

Forensic Volcanic Setting: Soufrière Hills Volcano, Montserrat

Montserrat is an active volcanic island within the Lesser Antilles volcanic arc (Figure 1). The Soufrière Hills volcano (SHV), located in the south of the Montserrat, became active in 1995 following a long period (estimated 400 years) of quiescence (Young et al. [1998]). This prompted an evacuation of the islands' capital city, Plymouth (located 4 km from the volcano summit; Figure 2i), and several nearby towns and villages. Following another intense phase of volcanic activity in 1997, many displaced Montserratians accepted a migration package to the UK and elsewhere in the Caribbean. A population of over 10,500 was reduced to just 2,850 (the population has since risen to 4,922 [2011 census]). The last significant activity occurred in February 2010, and while this is the longest pause in activity since 1995, it is not yet clear that the eruption has finished and is officially still on-going (Scientific Advisory Committee on Montserrat [2013]; Wadge et al. [2014b]).

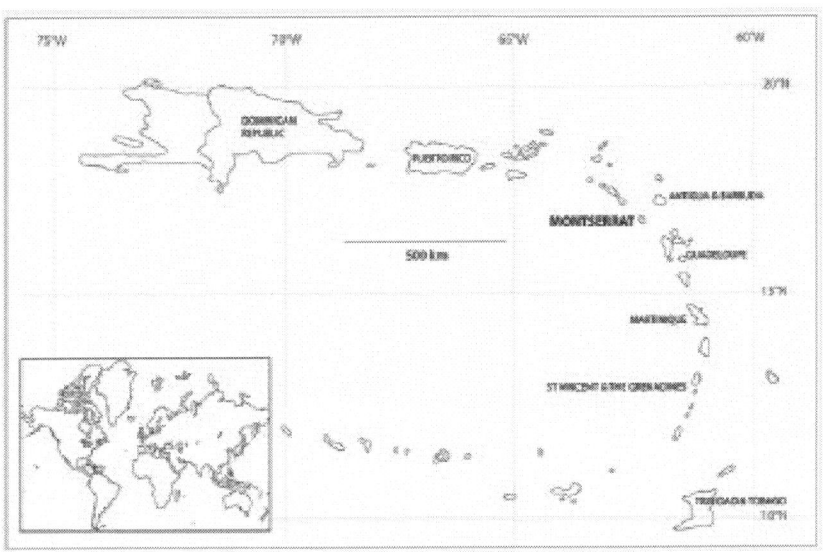

Figure 1: Sketch map of the Lesser Antilles, West Indies. Global position shown in the inset map.

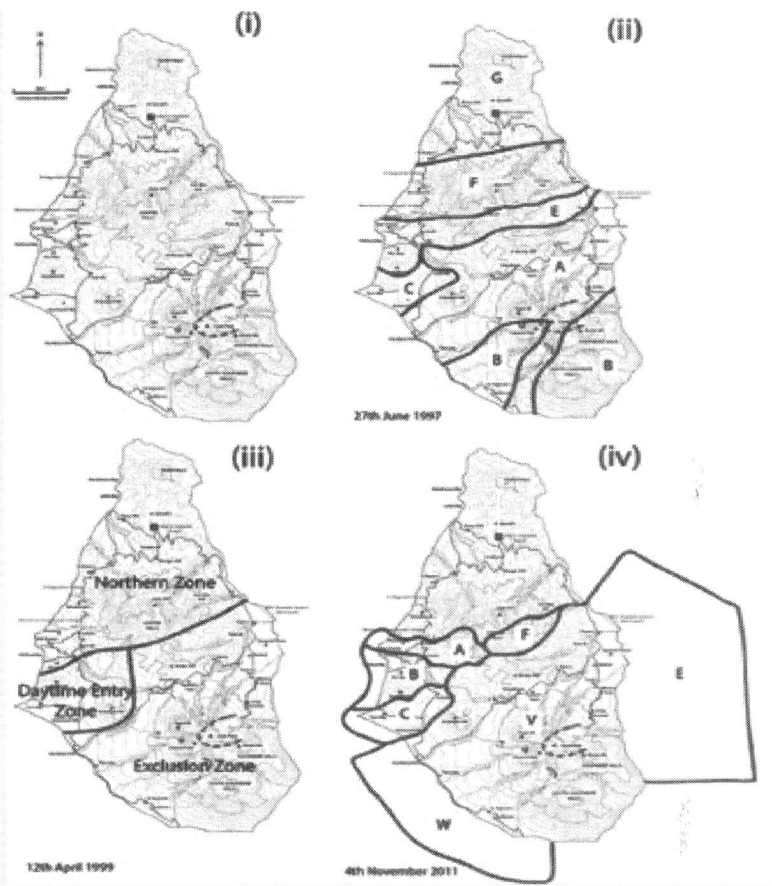

Figure 2: Maps of Montserrat showing major towns and cities, exclusion zones and major revisions to them over time. Map (ii) is one of several revised maps showing the microzonation of Montserrat into seven hazard zones (A-G). Access to some of these zones, particularly A and B, depended on the alert level (0 through to 5; 5 being the highest alert level rendering zones A-D inaccessible). Note that zone D was subsumed into zone C following the events of June 25th. Map (iii) shows the three broad zones which replaced microzonation. Map (iv) was implemented in August 2008 in response to the new hazard level systemhttp://www.mvo.ms/pub/Hazard_Level_System/ *webcite*. All maps have been redrawn from the original Montserrat Volcano Observatory maps.

The political links with the United Kingdom (Montserrat is one of 14 British Overseas Territories) and the long-lived nature of the eruption has rendered the Soufrière Hills Volcano one of the most well-studied in history. As expected, the rich literature resource is dominated by studies of the volcanic activity of SHV and, to a lesser extent, general economic, social, emotional, health, cultural impacts of the eruptive phase on Montserratian people and society (Halcrow Group Limited and the Montserrat National Assessment Team [2012]). However to date, there has been relatively little focus on the differentiation and dynamics of social vulnerability on Montserrat. This paper provides an analysis of vulnerable groups during the SHV crisis, and examines the processes of vulnerability generation.

We begin by describing our methodological approach to this study, framed around our 'forensic' approach. A description of the main volcanic phases of the SHV eruption follows, coupled with an account of some of the key social impacts during each phase. Finally we examine three of the most vulnerable groups that emerged from our analyses, followed by a discussion of social differentiation and dynamic vulnerability in volcanic settings.

METHODS

The STREVA project's modus operandum is to undertake interdisciplinary, detailed exploration of long-lived volcanic crises to significantly improve knowledge of the ways in which the components and drivers of volcanic risk interact and can be characterised, analysed and monitored. These explorations of the causes, impacts and trajectories of volcanic crises are termed 'forensic investigations' (Burton [2010]) and provide a platform for interdisciplinary teams to integrate systematic analyses of risk drivers, with a focus not just on the geophysical, but also on the wider societal drivers (e.g. governance, vulnerability, communication, infrastructure). The STREVA project focuses investigations around a forensic workshop, and combines this central data gathering activity with a series of key informant interviews and extensive study of a broad literature base. Each workshop is tailored for the particular context, but as a rule, they always include presentations, focus groups, and a field trip. The range of invited participants is also context-dependent, although crisis-response groups, government spokespersons and community representatives are always present. At each workshop, an event timeline is created by the workshop participants. This timeline records (on paper) physical and socially significant events before, during and after a volcanic crisis and provides a foundation upon which further multi-disciplinary data, gathered from other sources (i.e. interviews and literature), can be added and corroborated. Adopting a timeline-based approach provides a way of tracking events and impact pathways of the volcanic crisis on people and society, and illustrates responses and phases of change. In this paper, we present the results of this multi-phase data gathering approach, focused on analysing vulnerability during the SHV crisis.

The forensic investigation of the SHV crisis was centred around a two-day workshop, held in Montserrat, in September 2012. The aim of the workshop was to explore the extent to which Montserrat represents a resilient society by identifying the dynamic components and circumstances that have largely contributed to resilience, and those that have undermined it. Approximately 70 people attended the workshop, with scientists, government officials, disaster managers and community representatives in attendance. Involvement of the local community was a crucial component of the workshop, offering them an opportunity to have their voices heard, and to share individual and collective experiences and opinions. Workshops participants were carefully chosen to ensure that: a) numbers of scientists did not outweigh non-scientists; b) numbers of ex-patriates did not outweigh Montserratians; and c) most participants had been involved in, or had memory of, the SHV crisis.

The workshop was divided into two themes: the first was entitled, "what is resilience, where and when is it manifest on Montserrat and how is it controlled?" and the second was: "critical moments during SHV + 17[a]: dynamic risk, resilience and its drivers". Both sessions included presentations and facilitated break-out discussions, although for the latter theme, these group conversations were purposely designed to separate the delegates into particular 'specialisms': a) monitoring and hazards; b) risk and disaster managers and communicators; c) civil society. Each break out group was facilitated by a STREVA researcher with experience of these 'specialist' groups. The second theme was extended further by way of a series of charrettes to allow specialist groups to divide and re-form as mixed stakeholder groups. The final afternoon of the workshop was open to the public and included a panel session with key responders during the crisis.

[a]SHV + 17 = the 17 years since eruption onset (at the time of the workshop).

Each break out group was recorded and extensive notes were taken. These conversations provided a rich qualitative data set, which helped both to guide the choice of who to interview post-workshop and to inform the set of questions posed. Sixteen post-workshops interviews were conducted with members of the Montserratian population, government representatives, decision makers and the UK-based Montserratian community. These were almost always conducted in the office or home of the interviewee, lasted between 20–60 minutes, and were recorded and transcribed. Interviewees were chosen largely as result of 'knowledge gaps' identified from the workshop (e.g. returnees from the UK; immigrant population), or if they were representatives of vulnerable groups identified from the workshop conversations.

To analyse the vulnerability component of volcanic risk, the evidence gathered from the workshop and interviews, along with existing material (academic, grey and policy literature), were collated, triangulated[b] for validation and coded against a set of 14 impact and response themes. The themes of this outcome-based analysis[c] were selected for their consistency across the evidence base and included: ashfall; stress; evacuations; shelters; migration drivers; buffer zones; clean up; rebuilding services; and rebuilding livelihoods. A second phase of coding focused on vulnerable groups within these broad themes: original residents of the south; original residents of the north; shelter population (early in the crisis); shelter dependents (longer-term); relocatees to the north (home owners); relocatees to the north (renters); residents of buffer zones; migrants to the UK (non-assisted); migrants to the UK (assisted); migrants elsewhere; remittance receivers; returnees; and in-migrants. Again, these groups emerged from the data due to extent and regularity of references across the evidence base. We acknowledge that these are not uniform groups, but *for some* extended families which intersected several of these social groups, the combination of circumstances has exacerbated vulnerability and impaired recovery. Drawing on this outcome-based analysis, here we focus our narrative on three particularly vulnerable groups: shelter dependents (longer-term); relocatees to the north, and migrants to the UK (assisted).

[b]Triangulation refers to assessing, comparing and cross checking findings using a plurality of evidence sources, derived, for example, from diverse methods, informants, inquirers or contexts.

^cOutcome-based analysis of vulnerability, or end-point analysis, considers the impacts of volcanism (in this case) on specific social groups as an indicator (or manifestation) of underlying vulnerability, and employs social science research to explain the reasons why that vulnerability exists, varies and changes.

While it is recognised that rapid team-based qualitative inquiry can have limitations, particularly insensitivity to the social context and susceptibility to bias (Chambers [1994]), this was minimized in this case by: 1) encouraging a strategic mix of people to participate in the workshop (i.e. a combination of scientists and risk managers with detailed technical knowledge and experience of the SHV, and knowledgeable and representative local residents with direct experience of the crisis' longer-term impacts); 2) analysing group discussions to identify vulnerable groups, identify key informants and inform the design of subsequent semi-structured interviews; and 3) undertaking a series of subsequent in-depth interviews to deepen and substantiate the content of workshop discussions.

Before presenting results of our analyses, the following section briefly summarises the SHV crisis, the short term response, and its impact on the Montserratian population. While there have been five phases of volcanic activity since the crisis began, we have focused on the phases 1–3. The volcanological literature on the crisis is incredibly rich, and more comprehensive descriptions of events can be found elsewhere (e.g., Wadge et al. [2014a], [b]; Kokelaar [2002]; Loughlin et al.[2002]; Clay et al. [1999]).

THE SOUFRIÈRE HILLS VOLCANIC CRISIS: OVERVIEW OF IMPACTS AND RESPONSE

Phase One

On the 18th July 1995, volcanic activity of Soufrière Hills resumed after a long period of dormancy. The first large eruption occurred on the 21st August, known as 'Ash Monday', resulting in the evacuation of ~6000 people from Plymouth and nearby towns into temporary shelters (churches and schools). Evacuees reoccupied their properties two weeks afterwards, but volcanic activity temporarily forced them out again in December. On the 3rd April 1996, Plymouth was evacuated for the final time, and a state of public emergency was declared. Over 7,000 people had to be relocated, and 1,366 people were housed in temporary public shelters. Living conditions were widely viewed as unpleasant; evacuees complained about overcrowding and lack of privacy, poor sanitation, and lack of access to good nutrition. A voluntary evacuation scheme was set up on the 23rd April, offering Montserratians an opportunity to move to the United Kingdom, but only 1,244 people registered for this package. Several interviewees stated that this owed to hope that the eruption was short-lived. During this period, businesses began relocating to Brades, Salem and St John's, but towns in the North were struggling to hold any more evacuees, inducing further evacuations to the UK and elsewhere in the West Indies. In response to the atypical explosive-style of volcanism in September 1996, the island was 'microzoned' into seven hazard zones (A-G). The risk status of each zone was modified according to fluctuating alert levels, issued by the Government of Montserrat and based on recommendations from scientists at the MVO. Most areas in the north (zones G & F) could be fully occupied even when the alert level was at its highest, and most areas in the south (zones A & B) could not be accessed, or only accessed for short visits, when the volcanic dome was in a stable growth phase. All areas were accessible at the lowest alert level. Both the alert scheme and the maps were revised several times over the years (Figure 2), but these modifications resulted in some misunderstanding between the public, civil authorities and the scientists, and occasionally disregard to follow official warnings (Aspinall et al. [2002]).

On the 25th June 1997, a series of pyroclastic flows destroyed settlements and infrastructure (including the airport in the East) from Trants to Dyers and killed 19 people (Loughlin et al. [2002]). Salem, Old Towne and Frith were evacuated in August 1997 (zone E; Figure 2), forcing towns in the north to accommodate a further 1,300 people. Numbers in shelters rose from 775 to a peak of around 1,600 (Clay et al. [1999]). On the 19th August 1997, an assisted passage scheme was announced, which provided financial support for Montserratians to move to either the UK or a regional location. Over 4,000 people registered for the relocation package, and while some took advantage of pre-existing networks in the UK and moved in with friends and family (Shotte [2006]; McLeman [2011]), many were re-housed on estates in UK cities. On the 21st May 1998, the UK offered a permanent settlement deal to evacuees from Montserrat. This had considerable impact on demographics; early in 1998, the population of Montserrat was only 2,850 – a 70 % reduction from 10,625 (pre-eruption; 1991 census). On the 21st May 1998, the UK offered a permanent settlement deal to evacuees from Montserrat.

The risk map was simplified in September 1997 (final revision in April 1999), and microzones were replaced by three broad zones: exclusion, central and northern (Figure 2iii). This map remained largely the same until August 2008 (final revision November 2011) when the new hazard level system was implemented, in conjunction with a new hazard zone map, dividing the southern two-thirds of Montserrat into five zones (A, B, C, F & V) and two maritime exclusion zones (W & E) (Figure 2iv).

After the intense phase of activity waned, Montserrat began to rebuild, following an injection of funds from the Department for International Development (DfID), along with the creation of a sustainable development plan[d], and building of 'temporary' T-1_11 houses[e] in Davy Hill. This helped to reduce the numbers living in shelters to 427. In October 1998, the reoccupation of Salem, Old Towne and Frith began. Numbers of immigrants[f] began to rise to fill occupations left by evacuees. On the 1st May 1999, an assisted return passage scheme began, and as the population gradually increased to ~4,500, growth on the island was stimulated and the construction industry was re-established. A new housing development was built at Lookout in the North, further reducing the shelter population to 372.

[d]The first Sustainable Development Plan was devised in 1997 (covering the period 1997–2002) and was developed in the 2003–2007 document. The most recent plan, published in 2010, encompasses the period 2008–2020. All documents were developed by the Ministry of Economic Development and Trade, Government of Montserrat.

[e]While the 'temporary' housing is widely referred to as T1–11 housing, this is actually the code of the ribbed plywood siding itself. The housing units were timber framed with a plywood wall sheeting and a corrugated steel roof.

ᶠWhile exact numbers and nationalities of immigrants was unknown during this time, many interviewees reported that most immigrants originated from Santo Domingo and Guyana.

Phase Two

The volcano resumed dome growth in November 1999, beginning phase 2 (of 5) of the eruption. This was to be the longest phase of activity, pausing in July 2003. This phase was characterised by dome collapse events, sending pyroclastic flows down the Tar River Valley in the south-east (Figure 2). In October 2002 lahars affected the Lower Belham Valley area in the west (Figure 2). Residents were given 48 hours to evacuate. In the months that followed (296 days), residents were permitted to return to their homes in the exclusion zone between 09.00 and 14:00 (known as daytime entry), although access was withheld during periods of raised activity, due to the heightened risk of pyroclastic flows traveling down the valley. The Belham area was not permanently re-inhabited until after 13 July 2003 following a major dome collapse which greatly reduced risk to the Belham Valley.

Phase Three

In August 2005, a new lava dome began to develop, but it was not until the 20th May 2006 that explosions occurred, prefigured by the second largest dome collapse since the reactivation of SHV. Heavy ash falls affected most of the island, and required considerable investment from the Government of Montserrat to support further clean-up efforts. The necessity for extra manpower encouraged further immigration. The population according to the 2011 census was 4,922.

SHARPENING THE FOCUS: THE STORIES OF THE DISADVANTAGED

Results from the time-series analytical component of the forensic study are presented in Figure 3. Data used to inform this analysis was gathered from the forensic workshop, key informant interviews and available literature. Adopting a timeline-based approach provides a way of tracking events and impact pathways of the volcanic crisis on people and society, and illustrates responses and phases of change. While this impact timeline provides a useful illustration of the critical moments (and phases) in terms of social impacts on livelihoods and wellbeing to all Montserratians, vulnerability itself is a complex social characteristic and is more difficult to chart. While it can change, those changes are not necessarily sequenced by disaster events. However, the strength of impacts for different social groups can yield information on how vulnerability plays out, especially in the longer-term. Here we focus on trying to explain how particular social groups have proved to be more vulnerable than others, by analysing the impacts of the crisis on the 'worst affected' and the dynamics of that process. Essentially the timeline becomes a way to frame the vulnerability story of the disadvantaged, or those with the least capability to recover.

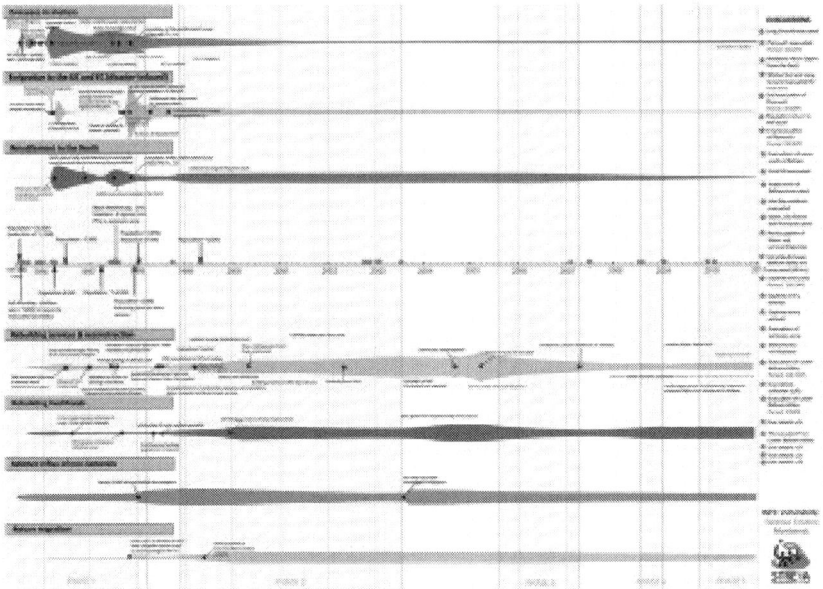

Figure 3: Annotated visualisation of the time-series analytical component of vulnerability. Major volcanic, social and political events are shown. Dates and details of evacuations are outlined on the right of the image. The five phases of volcanic activity are shown at the bottom of the image. While there is no scale (or x-axis) to this diagram, the 'lenses' signify population movements and economic growth.

Our analysis of the workshop and key informant data indicates that those most disadvantaged by the lasting impacts of the volcanic crisis were/are: evacuees in long-term shelter accommodation; poorer non-migrants who resettled in the north of Montserrat and assisted passage migrants to the UK. Most workshop participants and key informants suggested that this was due to these groups being in a prolonged vulnerable state as compared to other the broad groupings which emerged from the second coding phase. Here we present results from the experience and perspectives of members of these groups, and those who liaised directly with them, along with an explanation of the antecedent conditions that led to a more vulnerable position, the consequences and the long term situation.

Long-term Shelter Dependents

Those evacuees who remained in temporary shelters were predominantly families and individuals with fewer livelihood assets. There were two broad sub-groups of people: 1) those with lack of access to alternative accommodation via social networks in the north or other countries, and/or the economic means to rent or build new accommodation; and 2) older people that were left behind by migrant families.

Following the first evacuations in 1995, schools and churches were made into emergency shelters. Many people were encouraged to share houses with friends and family (families in the north were given allowances to house other people with them), but the numbers requiring the use of public shelters was still large, so tents were erected in Gerald's Park in the north of Montserrat (Clay et al.[1999]). In April 1996, metal prefabricated structures were erected in Brades, and timber chalets were erected for displaced people following the second evacuation of Plymouth. These could accommodate up to 20 people. Funding for emergency housing was not allocated until July 1997. As Clay et al ([1999]), p.34 report: "The public shelter programme was basically successful in providing everyone with immediate shelter, but was unsatisfactory in providing for more extended occupation." However, most shelter types were used for extended periods of time. For example, the metal shelters, which were reported to be unbearably hot to stay in during the day (Skelton [2003]), were still occupied three years after the onset of the crisis (Pattullo [2000]).

Several interviewees and focus groups reported that the consequences of shelter dependency included: exposure to health risks from poor sanitation, crowding and nutrition; risks to personal security; emotional stress/depression; and re-entry into to exclusion zones. In her account of the disaster, Pattullo ([2000]), p.91 describes the shelters as, "inadequate and often squalid". Personnel working at the Emergency Operations Centre (EOC), who were in charge of maintaining the shelters, found it particularly difficult to dispose of human waste:

"*The sanitation consisted of pit latrines less than 2 metres deep. They were intended to be in use for two days. They were, in fact, used for more than two years.*" (Pattullo [2000], p.94).

It is possible that this led to the increased levels of gastro-intestinal illness that was recorded during this period of the crisis. Many people were reliant on shelters, with up to 1,600 people (in August 1997) forced to tolerate crowded conditions, as the following quotes describe:

"*Now at the shelter you could imagine 20 persons living in this room after coming from a private home, just coping with that, just the emotional thought of that.*" (EOC Employee)

...*"we started in what we referred to as the rural areas, where you had the rural poor – you had a mixture of vulnerable groups, but clearly those were the ones who were impacted the most. Those are the ones that I saw in the shelters really suffer....Now a church is designed to have a two hour, at the most, celebration, and you return home. But when you convert that now to a place where folk are living, no partitions, and the restroom facilities were not designed for 60–100 people.*" (Community Services Employee)

The EOC were also in charge of food distribution, although provisions were mostly canned goods which were easier to disseminate. The quality of the food soon began to raise issues:

...*"because a lot of them were accustomed to eating a lot of what they produced – so that food situation was desperate for these vulnerable groups* [farmers]...*Right away, the stress of the situation, plus the food, created health problems.*" (Community Services Employee).

...*"we were seeing guys come in with foot and a half long carrots and we'd say hey, where was this grown and they'd say right up there* [the exclusion zone]... *the government agreed let us take some money and purchase some of the crops from them so that at least they could harvest and have some income and [for] the people in the shelter at least we could change the diet a little bit and make it a little easier...* (Community Services Employee).

Unfortunately this desire to help those in need, and to continue earning a living, encouraged some farmers to re-enter the exclusion zone.

"People died because they wanted to get crops to feed people in the shelters – the aid wasn't enough and people wanted to help the EOC feed people." (Government Official)

During the inquest[g] into the deaths of 25th June 1997, the jury decided that the failure of British and Montserratian governments to provide land for displaced farmers had contributed to the nine of the nineteen deaths.

[g]The inquest was held in November 1998, and the report published in January 1999.

If the presence of standing crops was a pull factor, so it appears that the conditions of shelter life were a push factor to re-enter the exclusion zone. Several workshop participants commented that some of those who died had either refused to move to the shelters or had returned to spend time at their homes in the exclusion zone. Two interviewees spoke not only of the physical deprivations of the shelters but also of personal security issues such as incidences of aggression, power struggles, rape and sexual abuse:

"...and to end up in a hall with 60 or how many people, [at night] people were being touched, and they are not certain who touching me...when we started there were no partitions – eventually an effort was made to use plywood and stuff and at least create some semblance." (Retired Community Services Employee)

There was some public disobedience fuelled by treatment in the shelters, and protests broke out (particularly in Salem). While this tension was ameliorated through the assisted passage scheme, conditions did not improve for those that stayed in shelters. As Clay et al. ([1999], p.33) report, "The conditions and length of time that people have had to endure living in public shelters have been regarded as unacceptable in terms of British and industrial country standards of social well-being."

One of the unanticipated negative consequences of families abandoning Montserrat was that many left their elderly on the island. A shelter for the elderly opened initially as many of the aging population required care as well as accommodation, but the insanitary and crowded conditions in this particular shelter drove several of the occupants to return to their homes in the exclusion zone (Loughlin et al. [2002]). There were reports of elderly becoming depressed and anxious (Avery [2003]; Stair and Pottinger [2005]), and in response, the government created three permanent residential homes.

There may have also been longer-term chronic health and mental health effects of extended shelter residence across the dependent population (Stair and Pottinger [2005]; Hincks et al. [2006]). While there is no concrete evidence for cause and effect, several focus groups claimed that poor diet (modest intake of fresh produce) and lack of exercise led to an increase in depression, hypertension, obesity and [symptoms of] diabetes. Today, almost 20 years after the onset of the crisis, there are still people occupying shelter accommodation (Sword-Daniels et al. [2014]), although exact numbers are uncertain.

Poorer Non-migrants who resettled in the North

Life for those outside the shelters was also challenging, especially for those struggling to establish homes and livelihoods following displacement from the south. As several focus groups participants and interviewees reported, many non-migrants who relocated to the north of Montserrat were initially reliant on the hospitality of family and friends or rented accommodation. This group included people who had never had substantial financial assets, but also home-owners from the south who were already paying mortgages for their abandoned homes (some people are still repaying loans on empty dwellings today). It was also due to shortage of available land:

"Land in north was family land so [they] did not want to sell it and there were difficulties getting the agreement of family members.... this partly explains why the government did not do more in the north, because it could not buy the land." (Government Official)

Further, the shortage of available land in the north meant that land prices rose sharply. While this had an effect on the Government of

Montserrat, who needed to purchase agricultural land from private landowners to start building housing developments, soaring prices particularly affected farmers, who could not afford to purchase replacement land:

> ... "land was at a premium here [in the north] because people wanted places to build houses, they wanted places to establish businesses...So the farmland started going at rates that were comparable to business leases." (Employee of the Department of Agriculture)

This further marginalised this rural social group. The destruction of farmland in the south meant that farmers lost their land, crops and livestock, and faced severe challenges in re-establishing their livelihood in the north (Rozdilsky, [2001]). Those that were able to rent land were faced with challenges of cultivating in unproductive, infertile soils on small plots, resulting in low yields. Consequently, few farmers could make a living solely from farming, so many had to obtain a second occupation such as fishing or construction (Halcrow Group and the Montserrat National Assessment Team [2012]). Some farmers who were able to transfer livestock from the south either did not have land sizeable enough for pasture, or were not able to build enclosures, so livestock were often unconfined. This in turn threatened crops of arable farmers. According to one interviewee, there were also reportedly instances in which tenant farmers were exploited, with landowners demanding the return of rented land, only after the land had been cleared by the tenants and started producing crops.

Overall, the farming sector of Montserrat has not only diminished but changed in trade from export to selling on-island and subsistence farming. A representative of the Department of Agriculture estimated there are just 70 farmers currently on Montserrat today, though only around five are full-time farmers, and some no more than 'backyard gardeners'. The longer-term situation of farmers renting in the north has also seen the diversification of farming methods and crops, apparently driven by in-migration from other islands:

> ...."we have a Haitian guy who is actually...renting some land and doing some serious farming, which I hope will inspire some of the other locals to get back in....The cuisine is changing because of people of different [cultures], and so in farming you have crops like pak choi.... Those kind of things you find people now planting that and selling that – some of the locals even planting some of these different crops" (Employee of the Department of Agriculture)

However, as emphasised by one interviewee, others have not been keen to take risks in their approaches to farming, as they do not have security of tenure, and are aware of the threat of future ashfall and acid rain (Halcrow Group and the Montserrat National Assessment Team [2012]). The risk of acid rain and heavy ashing was persistent throughout the post-evacuation period, when a change in wind direction would re-direct the sulphurous gas/ash plume over to the north, sometimes causing crops to disappear overnight.

Over time, new houses for relocatees were built in the north, but several interviewees raised concerns about the structural quality and location of some of this housing. One of the longer term effects of the land shortage in the north was that some new homes (even government-funded housing estates) were erected in unsafe and unsuitable locations such as ravines. One of the new housing developments, Lookout, comprised of 200 units, eventually enabled many renters to purchase their properties. However, some houses in the development were built in apparently poorly planned locations – on steep, exposed slopes that have limited shelter against hurricanes, earthquakes and windborne salt (Mitchell [2001]; Smith Warner Report [2003]). Another re-development in Little Bay is also at risk from regular coastal flooding (Mitchell [2001]). Possibly due to the lack of available space on which to re-build and/or the time pressures of relocating shelter dependents, hazard mitigation did not play a central role in the initial re-building phase, thus further affecting the vulnerability of these affected groups.

Assisted Passage Migrants to the UK

The motivation to register for the relocation package was a major subject of discussion in the workshop focus groups as well as in several of the key informant interviews. The balance of perspectives was that motivation for those evacuated from the south arose from a combination of necessity and opportunity. Strong drivers were the dependence on shelters and mortgage commitments of abandoned homes:

"One of the big failures at the beginning of the volcanic crisis was that the insurance companies all closed and they didn't pay people out....a lot of people had mortgages...that's one of the reasons people left who otherwise would have stayed, because their paychecks were docked for the mortgage payments [on abandoned homes]" (Workshop participant [Montserratian resident])

However, it was not only the relocatees from the south who registered for the package:

"The other thing we spotted was the some people in the north decided to move because they were just so overwhelmed by so many people coming into their environment. Some just locked their houses and took off." (Retired community services employee)

... *"More emigrated from the north as they mostly worked in agriculture and had no employment post-eruption, whereas people from the south were more likely to be in government jobs which were maintained."* (Workshop participant [Montserratian])

This did have some positive consequences for the distribution of the remaining population:

"The fortunate thing is that...for some of us, is that some of the people from the north, who did not necessarily have to migrate, wanted the chance to go to the UK, so that migrated and they created space for some of us [people from the south]" (Employee Department of Agriculture)

Many reported on bring fearful of the eruption and the potential for larger, on-going activity. Reports on health issues during the eruption also underlined that migrants were motivated by health concerns from inhaling ash and from the poor hygiene in the shelters, particularly for children (Avery [2003]; Forbes et al. [2003]; Howe [2003]). A study by Forbes et al ([2003]), conducted in 1998, reported that, "children who lived in areas with moderate or heavy exposure to ash since July 1995 reported more respiratory symptoms and use of health services for respiratory problems than children who had never lived in these areas" (p.209). Further, she reports, "asthma was frequently cited as a medical condition among families who left under the Assisted Passage Scheme" (p. 209).

People were also incentivized to take up the package to the UK, by housing offers and financial support, and access to employment, health services, and schooling. The drive to seek educational opportunities for children was one of the most common themes in the discussions. However in many cases, this led mothers and children alone to migrate (Young [2004]; Shotte [2007]; McLeman[2011]). This can be articulated as both a pull and a push factor, in that the education system in the island was initially severely disrupted by loss of facilities, departure of staff and by the usage of schools in the north as shelters.

The separation of family members was one of the major consequences of the migration process, as in many cases, fathers and grandparents were left behind:

"….so at one stage we had the men here who were basically involved in the reconstruction and trying to keep things going, but the partners were in England and in different parts of the Caribbean with the children. So that created another social problem, in that some of the children adapted well, but some of them, exposed to a new environment, just couldn't handle it and we heard about them getting into trouble." (Retired community services employee)

Given that many families were forced to relocate without their fathers, this lack of paternal support may have influenced the behaviour of some students (Shotte [2002]). There were some accounts of poor performance in schools, reportedly as a result of 'corrupting influences' of some British school children, and/or the challenges of students trying to defend their ethnic identity (Shotte [2006]). Whatever the reason for perceived, or actual behaviour change, several students clearly struggled with the challenges of adapting to a new social and cultural situation: "the seeking of autonomy and independence from parents, together with constant identity reconstruction have put extraordinary emotional strain on relocated students psyches' – a situation that has impacted negatively on their overall educational progress" (Shotte [2006], p.34). Despite better education being a driving force for emigration, there were also reports of some Montserratian parents' displeasure at some teaching standards and the perceived regression of learning (Windrass and Nunes [2003]); a perception rooted perhaps in the difference in relationships between teaching and learning in different cultures (Montserrat as an interdependent culture and Britain as an independent one).

Particularly for those migrants without access to pre-existing networks with friends and family in the UK, the new conditions they faced posed several challenges, especially in the early years of the crisis (Shotte [2007]). Migrants were dispersed to several UK cities, and many were re-housed in socially-deprived neighbourhoods and hostels. They also had to cope with the challenges of acquainting themselves with unfamiliar economic, social and cultural situations. Some appear to have coped well, but others struggled severely in the early years. In particular, there were challenges of benefit delay, attributed to the difficulties of obtaining a National Insurance number (Pattullo [2000]). Further, some rented accommodation also lacked furnishings, including beds. There were challenges of finding employment without references or recognised qualifications, even for those who had positions of accountability in Montserrat (Pattullo [2000]). For many older people who did make the move under the assisted migration package, re-establishing life in the UK appears to have been particularly difficult:

"Depending on where they landed we got reports back that some did fare very well in terms of the organization and them learning the system, which would have been difficult for the real old persons without family members. And we just kept hearing the announcements of a number of them in those early days that they got there but what next, what is the will to live. So ....from going outside and planting your garden and harvesting your stuff to now locked up in an apartment. Granted you are getting the dole but these people are not dole people. And it's cold. These people are people who are 70 odd, 80 years of age and they got up every morning and tied out their animals and planted some food, cooked their food. Now you took them into an apartment and you say, ok, you don't have to worry you'll get your food etc – but for some it was a real shock." (Retired community services employee)

In the longer-term, the UK-based Montserrat community have maintained their cultural and emotional ties to Montserrat (Shotte [2007]; Hill [2014]) but there are low levels of return - only 60 people took up the return package to Montserrat when it was offered in 2003. Montserratians are unlikely to return to the island until more of the island becomes accessible, and employment and housing opportunities are comparable to the UK:

"Lots of Montserratians would think twice about going back to invest. In terms of setting up a big business [you] have got to look at population." (UK-based Montserratian)

However it is not just population numbers that seem to discourage migrants to return to Montserrat; the present-day population structure is such that there is also an anxiety that Montserratian culture has largely been lost (Greenaway [2011]; Hill [2014]). Further, it is also likely that long-term residency in the UK has created an inter-generational adaptability to the British way of life:

"The majority of people in the UK are settled. Some people who wouldn't have had a chance to go to university in Montserrat - they

*have the knowledge but not the cash – but in the UK they do. ....
[The] majority of people make use of the opportunity.* (UK-based
Montserratian).

DIFFERENTIATED AND DYNAMIC VULNERABILITY

The story of Montserrat's volcanic eruptions received several pages of
coverage in the second edition of *At Risk*, a landmark publication in
the evolution of ideas around disaster risk and social dimensions of
vulnerability (Wisner et al. [2004]). Though most of the discussion,
as elsewhere, focussed on island-wide risk factors and the overall
management of the crisis, the authors did make reference to a social
differentiation in how the short and medium-term consequences
played out for different social groups. They begin this passage with the
following words:

'Volcanoes can be seen as great levellers, potentially deadly to all
life and all people, rich or poor, who are within reach of their destructive
power. But in the case of Montserrat there were exceptions.' (Wisner
et al [2004], p.307).

It is the story of the 'exceptions' that forms the focus of this paper.
In doing so, we can question just how exceptional these 'exceptions'
are in the context of volcanic risk.

Discussion of risk associated with volcanoes has generally tended
to focus on immediate threats to life from exposure to lethal volcanic
hazards. Such losses of life were kept numerically low in Montserrat by
the successive evacuations, but 19 people were killed by pyroclastic
flows on 25th June 1997. The return of people, mostly smallholder
farmers, to the flanks of the volcano contributed to this loss of life. It
has been argued that, for many, a combination of the strains of shelter
life, shortage of cash and the fear of losing standing crops and livestock
was motivation to return (Wisner et al. [2004]). This was a particular
social group facing difficult living conditions in the public shelters and
with high livelihood dependence on land that they could not replace
in the north.

However, as with other volcanic disasters, the vast majority of people
severely affected by the Montserrat eruptions were well out of the way

when lethal flows swept down the valleys from Soufrière Hills. For the surviving majority it is vulnerability in terms of ongoing livelihood and wellbeing that becomes key, and in Montserrat social differences in underlying vulnerability become manifest in the unfolding story of crisis and response. These differences became especially heightened by the longer-term impacts of the crisis – beyond the immediate effects of the hazard events.

Critical social science research on disaster risk emphasizes the importance of pre-existing assets and resources in shaping both how seriously affected people's lives may be by hazards and how readily they may recover (Anderson and Woodrow [1998]; Chhotray and Few [2012]). Vulnerability to hazards is therefore inherent in antecedent conditions. The preceding section has described the difficult situation faced by long-term shelter dependents, poorer non-migrants and assisted passage migrants in the years following the major eruptions. These groups are not discrete and there are overlaps and linkages between them that reinforced the social pattern of impact, for example, in the situation faced by older people left behind as other family members took assisted passage, and in the eventual movement of people from shelter to poor quality housing or rental accommodation.

The volcano's impacts touched everybody, and the hardship was widely distributed, as people lost their homes and businesses in the south. But many displaced residents had the financial or social means to bypass the shelters (or move rapidly from them) to alternative homes in the north or overseas, and were able to access savings, credit or assistance from personal networks to rebuild their standards of living. By contrast, many of those most vulnerable during and after the major eruptive phase came into the crisis with lower incomes, fewer economic assets and limited social networks (e.g. marginalised farmers not living in the former capital Plymouth). In terms of impacts of the crisis on these groups, relatively fewer livelihood assets constrained options for accommodation, migration and occupation. This parallels wider work on household decision-making in development studies that highlights how constrained access to livelihood assets limits the adaptive choices available to people in response to shocks and stresses (e.g. Ellis [2000]; McDowell and Hess [2012]). The outcome for these groups was heightened deprivation extending beyond the immediate impacts of the eruption and evacuation of the south into the medium-term.

In other words, if applying mainstream disaster management concepts, one can see that severe impacts of the hazards were experienced for these groups for several years at least into the disaster 'recovery' period. Their differential vulnerability to the volcanic hazards therefore became manifest through the unfolding of the volcanic-generated social crisis that ensued – in ways that could not simply be 'read off' from the physical effects of the ashfalls and the pyroclastic flows.

While perhaps this reflects a predictable pattern of underlying vulnerability, shaped largely by pre-existing relative poverty and/or social marginalization it is essential also to reflect on the dynamics of vulnerability, and how those patterns may have shifted during the course of the crisis (see Rigg et al.[2008] for an exploration of equivalent vulnerability dynamics following the 2004 Indian Ocean tsunami). Though we did not have the chance to collect data to verify this, it is possible that others moved into a condition of poverty and marginalization through loss of homes and businesses in the south and withdrawal of insurance cover, and through dispersion of social networks through the displacement and migration process. Hence people may have experienced impacts that subsequently undermined their pre-existing capacities to cope and manage crisis – they effectively joined the highly vulnerable group through the passage of events.

Certain crisis management decisions, actions and inactions - some of these associated with the political difficulties of working in a situation of high uncertainty - were key in shaping vulnerability dynamics (see Clay et al. [1999]; Wilkinson [2015] [in this volume]). Some such actions may have been unavoidable. For example, the major evacuations (and smaller, repeated ones) and the exclusion zoning led to a loss of non-fixed assets such as crops and livestock, as well as fixed assets. Inability to retrieve those assets impaired recovery. However, we have seen that the enforcement of exclusion was not entirely complete, which, on the one hand enabled some people to attempt to retrieve assets, and, on the other hand, endangered their lives.

Other aspects of crisis management which may have been handled differently also shaped shifts in vulnerability. The poor shelter conditions appear to have prompted several people to return to homes and landholdings in the exclusion zone, and thus increased their hazard exposure. Slow progress in establishing land and homes for resettlement in the north prolonged the time in which people with limited alternatives had to reside in inadequate shelter. The nature of migrant placements in the UK radically transformed the social context for the migrants, and insufficient support for families in difficult environments without ready access to social networks appears to have exacerbated emotional and behavioural problems and contributed to educational issues.

However, this situation was not *necessarily* permanent nor an inter-generational 'trap'. Just as people's circumstances can change negatively so they can change for the positive. The story of Montserrat's volcanic crisis is one of constant flux – physically and socially – and over the longer term, a form of recovery has set in, even though the cultural and environmental landscape in which it is doing so has been radically altered. Those who remained, moved to, and returned to Montserrat, have the chance of access to improved housing schemes and are seeing business and economic opportunities being rebuilt since the devastation of the south of the island. Many of the assisted migrant families who remained in the UK have experienced social mobility over time, particularly as younger members began to emerge from education into a wider job market than existed in Montserrat. Disasters can have the potential to act as moments of wider social change (Pelling and Dill [2010]) – though we would always argue that vulnerability analysis requires us to look beyond the aggregate to see how both impacts and recovery trajectories are socially differentiated.

CONCLUSIONS

Are volcanoes 'great levellers', or should we indeed expect their effects, like most other natural hazards, to be far from even? Evidence from Montserrat, and also from research at other volcanoes, such as Pinatubo (Crittenden et al [2003]; Gaillard [2008]), suggest that we should expect the latter. To be sure, the chances of surviving direct exposure to a pyroclastic flow are close to zero, whoever you may happen to be. But, the chances of coping through the ensuing disruption, of maintaining wellbeing and of recovering losses and rebuilding livelihoods are highly variable, shaped both by individual characteristics and by social structures.

Vulnerability to hazards is a complex and socially differentiated characteristic. The differentiation of effects is especially manifest over the long-term in a prolonged crisis and one involving radical disruption, as in Montserrat. Reports to date have generally discussed overall impacts and disaster management, and there is much ongoing debate about the resilience of the general island population. But we also need to look beyond the general context to ask who's impacts, who's recovery, who's resilience? Using mixed sources, the forensic work on Montserrat was an opportunity to collate the stories of the most vulnerable groups - and view how the medium/long-term impacts of the volcanic crisis on these social groups were linked to large extent with preceding socio-economic conditions.

However, it is also important to understand the dynamics of vulnerability, particularly through the course of a long-lived crisis. In an individual sense, people's lives were in flux through the duration of the crisis – some lost, some gained key assets that changed the nature of their vulnerability to ongoing impacts. In a wider sense, physical events and organizational decisions and inactions actively accentuated the social differentiation of impacts, through the processes of evacuation, shelter provision, resettlement, rehabilitation and migration. Because vulnerability is shaped by so many interlocking social and environmental factors change in vulnerability is not necessarily *sequenced* by disaster events. However, successive hazards and the variable responses they can trigger do constitute a dynamic that on occasion can be 'game-changing'. By taking a partial view of causation here – we can see how the specific unfolding of events in Montserrat led to social outcomes (or manifestations of vulnerability) that were not entirely predictable when the emergency began.

AUTHORS' CONTRIBUTIONS

AH and RF both attended the STREVA forensic workshop in Montserrat and gathered focus group and key informant data. Coding and analyses were conducted by AH and RF. AH and RF drafted the manuscript. Both authors read and approved the final manuscript.

ACKNOWLEDGMENTS

The authors wish to express their sincere gratitude to all those from Montserrat who participated with enthusiasm and interest in the forensic research process. We also acknowledge the support given by Jenni Barclay and the wider STREVA team, particularly: Paul and Liz Cole for facilitating access on Montserrat; Emily Wilkinson and Jonathan Stone for contributing to data acquisition; and Peter Simmons, Richard Herd and Jenni Barclay for contributing to the data coding process. The authors wish to thank Steve Sparks for an internal review of this manuscript, Victoria Sword-Daniels for helpful remarks, Chris Kilburn for editorial control, and two anonymous reviewers for comments on an earlier version of this paper. Funding for the STREVA project is provided by the NERC-ESRC Increasing Resilience to Natural Hazards Programme (grant number NE/J020052/1). Financial support for this open-access manuscript was kindly provided by the University of East Anglia.

REFERENCES

1. Anderson MB, Woodrow PJ (1998) Rising from the ashes: development strategies in times of disaster. IT Publications, London.

2. Aspinall WP, Loughlin SC, Michael FV, Miller AD, Norton GE, Rowley KC, Sparks RSJ, Young SR (2002) The Montserrat volcano observatory: its evolution, organization, role and activities. In Druitt TH, Kokelaar BP (eds) The eruption of Soufrière Hills Volcano, Montserrat, from 1995 to 1999. Geological Society, London, Memoirs 21(1):71-91

3. Avery JG (2003) The aftermath of a disaster: recovery following the volcanic eruptions in Montserrat, West Indies. West Indian Med J 52:131-135

4. (2004) Mapping vulnerability: disasters, development and people. Earthscan, London.

5. Burton I (2010) Forensic disaster investigations in depth: a new case study model. Environment 52(5):36-41

6.   Chambers R (1994) The origins and practice of participatory rural appraisal. World Dev 22(7):953-969

7.   Chhotray V, Few R (2012) Post-disaster recovery and 'ongoing' vulnerability: ten years after the super-cyclone of 1999 in Orissa, India. Glob Environ Chang 22:695-702

8.   Clay E, Barrow C, Benson C, Dempster J, Kokelaar P, Pillai N, Seaman J (1999) An evaluation of HMG's response to the Montserrat volcanic emergency, 2 vols. Evaluation report EV635. Department for International Development, London.

9.   Crittenden KS, Lamug CB, Nelson GL (2003) Socioeconomic influences on livelihood recovery of Filipino families experiencing recurrent lahars. Philipp Sociol Rev 51:115-134

10.   Cutter SL (1996) Vulnerability to environmental hazards. Prog Hum Geogr 20:529-539

11.   Dibben C, Chester DK (1999) Human vulnerability in volcanic environments: the case of Furnas, Sao Miguel, Azores. J Volcanol Geotherm Res 92(1–2):133-150

12.   Ellis F (2000) The determinants of rural livelihood diversification in developing countries. J Agric Econ 51(2):289-302

13.   Few R (2007) Health and climatic hazards: framing social research on vulnerability, response and adaptation. Glob Environ Chang 17(2):281-295

14.   Forbes L, Jarvis D, Potts J, Baxter PJ (2003) Volcanic ash and respiratory symptoms in children on the island of Montserrat, British West Indies. Occup Environ Med 60:207-211 doi:10.1136/oem.60.3.207

15.   Gaillard JC (2008) Alternative paradigms of volcanic risk perception: the case of Mt Pinatubo in the Philippines. J Volcanol Geotherm Res 172(3–4):315-328

16.   (2011) Montserrat in England: dynamics of culture. iUniverse, Bloomington IL.

17.   Halcrow Group Limited and the Montserrat National Assessment Team (2012) Country Poverty Assessments vol. 2. Supplementary Material pp.162.

18.   Hicks A, Barclay J, Simmons P, Loughlin S (2014) An interdisciplinary approach to volcanic risk reduction under conditions of uncertainty: a case study of Tristan da Cunha

Nat. Hazards Earth Syst Sci 14:1871-1887 doi: 10.5194/
nhess–14–1871–2014

19. Hill L (2014) Life after the volcano: the embodiment of small
    island memories and efforts to keep Montserratian culture alive
    in Preston, UK. Area 46(2):146-153 doi: 10.1111/area.12084

20. Hincks T, Aspinall W, Baxter P, Searl A, Sparks R, Woo G (2006)
    Long term exposure to respirable volcanic ash on Montserrat: a
    time series simulation. Bull Volcanol 68(3):266-284

21. Howe T (2003) The impact of the Montserrat volcanic eruption on
    water and sanitation 1995–1997, and lessons learned. Prepared
    for the 12th Annual CWWA Conference, Bahamas pp. 12 www.
    bvsde.paho.org/bvsacd/cwwa/howe.pdf.

22. Kokelaar BP (2002) Setting, chronology and consequences of
    the eruption of Soufrière Hills Volcano, Montserrat (1995–1999).
    In Druitt TH, Kokelaar BP (eds) The eruption of Soufrière Hills
    Volcano, Montserrat, from 1995 to 1999. Geological Society,
    London, Memoirs 21(1):1-43

23. Loughlin SC, Baxter PJ, Aspinall WP, Darroux B, Harford CL, Miller
    AD (2002) Eyewitness accounts of the 25 June 1997 pyroclastic
    flows and surges at Soufrière Hills Volcano, Montserrat, and
    implications for disaster mitigation. In Druitt TH, Kokelaar BP (eds)
    The Eruption of Soufrière Hills Volcano, Montserrat, from 1995 to
    1999. Geological Society Memoir 21(1):211-230 doi:10.1144/
    GSL.MEM.2002.021.01.10

24. McDowell J, Hess J (2012) Accessing adaptation: multiple
    stressors on livelihoods in the Bolivian highlands under a
    changing climate. Glob Environ Chang 22:342-353

25. McLeman RA (2011) Settlement abandonment in the context of
    global environmental change. Glob Environ Chang 21(suppl.
    1):S108-S120

26. Mitchell T (2001) Discussion of "Second hazards assessment and
    sustainable hazards mitigation: disaster recovery on Montserrat"
    by Rozdilsky R. Natural Hazards Review 74(2,2):64-71

27. Pattullo P (2000) Fire from the mountain: the tragedy of Montserrat
    and the betrayal of its people. Constable, London.

28. Pelling M (2003) Disaster risk and development planning: the
    case for integration. Int Dev Plan Rev 25(4):i-ix

29. Pelling M, Dill C (2010) Disaster politics: tipping points for change in the adaptation of socio-political regimes. Prog Hum Geogr 34:21-37

30. Rigg J, Grundy-Warr C, Law L, Tan-Mullins M (2008) Grounding a natural disaster: Thailand and the 2004 tsunami. Asia Pacific Viewpoint 49(2):137-154

31. Rozdilsky JL (2001) Second hazards assessment and sustainable hazards mitigation: disaster recovery on Montserrat. Natural Hazards Review May 2001:64-71

32. Schipper L, Pelling M (2006) Disaster risk, climate change and international development: scope for, and challenges to, integration. Disasters 30(1):19-38

33. Scientific Advisory Committee on Montserrat (2013) Assessment of the hazards and risks associated with the Soufrière Hills Volcano, Montserrat. Eighteenth report of the scientific advisory committee on Montserrat Volcanic Activity. Montserrat Volcano Observatory Part II: Full Report pp. 76 www.mvo.ms/pub/SAC_Reports/SAC18-Full.pdf.

34. Shotte G (2002) Education, migration and identities: relocated Montserratian secondary school students in London schools. Ph.D. thesis, Institute of Education, University of London.

35. Shotte G (2006) Identity, ethnicity and school experiences: relocated Montserratian students in British schools. Refuge 23(1):27-39

36. Shotte G (2007) Diasporic transnationalism: relocated Montserratians in the UK. Caribbean Quarterly 53(3):41-69

37. Skelton T (2003) Globalizing forces and natural disaster: what can be the future for the small Caribean island of Montserrat? In: Kofman E, Young G (eds) Globalization: theory and practice, 3rd edn. Continuum, London.

38. Smith Warner International (2003) Integrated vulnerability assessment of Montserrat. Submitted to the Government of Montserrat, June 2003 pp.76.

39. Stair AG, Pottinger AM (2005) Disaster preparedness and management in the Caribbean: the need for psychological support. West Indian Med J 54(3):165-166

40. Sword-Daniels V (2011) Living with volcanic risk: the consequences of, and response to, ongoing volcanic ashfall from a social infrastructure systems perspective on Montserrat. N Z J Psychol 40(4):131-138

41. Sword-Daniels V, Wilson TM, Sargeant S, Rossetto T, Twigg J, Johnston DM, Loughlin SC, Cole PD (2014) Consequences of long-term volcanic activity for essential services in Montserrat: challenges, adaptations and resilience. In Wadge G, Robertson REA, Voight B (eds) The Eruption of Soufrière Hills Volcano, Montserrat from 2000 to 2010. Geological Society, London, Memoirs 39:471-488 doi:10.1144/M39.26

42. Wadge G, Robertson REA, Voight B (2014) The eruption of Soufrière Hills Volcano, Montserrat from 2000 to 2010. Geological Society, London, Memoirs 39:pp501

43. Wadge G, Voight B, Sparks RSJ, Cole PD, Loughlin SC, Robertson REA (2014) An overview of the eruption of Soufrière Hills Volcano, Montserrat form 2000 to 2010. In Wadge G, Robertson REA, Voight B (eds) The Eruption of Soufrière Hills Volcano, Montserrat from 2000 to 2010. Geological Society, London, Memoirs 39:1-40 doi: 10.1144/M39.1

44. Wilkinson E (2015) Beyond the volcanic crisis: co-governance of risk in Montserrat. Journal of Applied Volcanology 4(3) doi:10.1186/s13617-014-0021-7.

45. Windrass G, Nunes T (2003) Montserratian mothers' and English teachers' perceptions of teaching and learning. Cogn De v 18(4):555-577

46. Wisner B, Blaikie P, Cannon T, Davis I (2004) At risk: natural hazards, people's vulnerability and disasters. Routledge, London.

47. Young I (2004) Monserrat: post volcano reconstruction and rehabiliation–a case study. Proceedings of the second international conference on post-disaster reconstruction: planning for reconstruction, Coventry university UK. 22-23

48. Young SR, Sparks RSJ, Aspinall WP, Lynch LL, Miller AD, Robertson REA, Shepherd JB (1998) Overview of the eruption of Soufrière Hills Volcano, Montserrat, 18 July 1995 to December 1997. Geophys Res Lett 25(18):3389-3392

Variations in Community Exposure to Lahar Hazards from Multiple Volcanoes in Washington State (USA)

Angela K Diefenbach[1], Nathan J Wood[2], and John W Ewert[1]

[1]U.S. Geological Survey, Cascades Volcano Observatory, 1300 SE Cardinal Court, Bldg 10, Suite 100, Vancouver 98683, WA, USA

[2]U.S. Geological Survey Western Geographic Science Center, 2130 SW 5th Ave., Portland 97201, OR, USA

ABSTRACT

Understanding how communities are vulnerable to lahar hazards provides critical input for effective design and implementation of volcano hazard preparedness and mitigation strategies. Past vulnerability assessments have focused largely on hazards posed by a single volcano, even though communities and officials in many parts of the world must plan for and contend with hazards associated with multiple volcanoes. To better understand community vulnerability in regions with multiple volcanic threats, we characterize and compare variations in community exposure to lahar hazards associated with five active volcanoes in Washington State, USA—Mount Baker, Glacier Peak, Mount Rainier, Mount Adams and Mount St. Helens— each having the potential to generate catastrophic lahars that could strike communities tens of kilometers downstream. We use geospatial datasets that represent various population indicators (e.g., land cover, residents, employees, tourists) along with mapped lahar-hazard boundaries at each volcano to determine the distributions of populations within communities that occupy lahar-prone areas. We estimate that Washington lahar-hazard zones collectively contain 191,555 residents, 108,719 employees, 433 public venues that attract visitors, and 354 dependent-care facilities that house individuals that will need assistance to evacuate. We find that population exposure varies considerably across the State both in type (e.g., residential, tourist, employee) and distribution of people (e.g., urban to rural). We develop composite lahar-exposure indices to identify communities most at-risk and communities throughout the State who share common issues of vulnerability to lahar-hazards. We find that although lahars are a regional hazard that will impact communities in different ways there are commonalities in community exposure across multiple volcanoes. Results will aid emergency managers, local officials, and the public in educating at-risk populations and developing preparedness, mitigation, and recovery plans within and across communities.

INTRODUCTION

The catastrophic destruction associated with recent lahars (e.g., 1985 Nevado del Ruiz, Colombia; 1991 Mount Pinatubo, Philippines; 1998 Casita, Nicaragua) has raised global awareness of this ground-based volcanic hazard. Lahars, which are high-concentration mixtures containing water and solid particles of rock, ice, wood, and other debris, are significant volcanic hazards to downstream communities because of the fast speeds and the long distances they can travel from their source. Between AD 1600 and 2010, lahars triggered during volcanic eruptions killed 37,451 people worldwide, including 23,080 in the 1985 Nevado del Ruiz disaster alone (Witham [2005]; Auker et al. [2013]). Although typically triggered during or shortly after volcanic eruptions, lahars can also be initiated by noneruptive events, such as heavy precipitation, earthquakes, and gravitational failure, making it difficult to forecast their occurrence (Pierson [1989]; Scott et al. [2001]). Regardless of how they have formed, recent lahars have caused major economic and business interruption losses, destruction of property and infrastructure, and tragic loss of life (Tayag and Punongbayan [1994]; Voight [1996]; Scott et al. [2005]).

Throughout the world, various approaches to reduce societal risks associated with lahar hazards have been implemented in communities downstream of volcanoes (see Pierson et al. [2014] for a review). Because preparedness and mitigation resources are not limitless, societal vulnerability assessments are increasingly being used as tools for targeting and prioritizing risk-reduction resources (Birkmann [2006]). A clear understanding of how a system is specifically vulnerable to a hazard (typically described in terms of exposure, sensitivity, and adaptive capacity) can help emergency managers and local officials identify opportunities for preparing at-risk communities and mitigating potential losses (Wood [2009]; Wood [2011]). Knowledge of the number, characteristics, and distribution of people exposed to a hazard provides insight on where potential losses could be the greatest, where potential challenges may exist in responding to and recovering from hazardous events, and the underlying factors that create and amplify societal vulnerability to hazards (National Research Council [2012]). In addition, individuals that understand the potential impacts of a hazard in their community and how they are specifically vulnerable are more likely to be involved in planning efforts and pro-active in preparedness strategies (Paton et al. [2001]; National Research Council[2012]).

Research related to societal vulnerability to volcano hazards has focused on various aspects, such as exposure metrics of people and infrastructure (Rapicetta and Zanon [2009]; Wood and Soulard[2009a], [b]; Kunzler et al. [2012]), community resilience (Paton et al. [2001]; Tobin and Whiteford[2002]), individual risk perceptions (Lavigne et al. [2008]; Johannesdottir and Gisladottir [2010]), and community preparedness (Gregg et al. [2004]). Vulnerability has been assessed using different methods, ranging from geospatial overlays of hazards and assets (Aceves-Quesada et al. [2007]) to probabilistic loss assessments (Spence et al. [2005]), and from municipal (Kaye et al. [2009]) to global (Chester et al. [2000]) scales.

Although past vulnerability assessments provide valuable insight to public officials and emergency managers, they have typically focused on individual volcanoes. However, in many parts of the world, emergency managers must contend with and plan for multiple volcanoes in their jurisdictions. Lahar mitigation-decisions then become a difficult process given the multiple sources, the spatial extent of potential threats, the multiple jurisdictions, and the various elements of societal vulnerability (e.g., population, economic, infrastructure, and environmental assets). Therefore, practitioners with resource limitations need methods for determining and prioritizing on which lahar hazards to focus their limited risk-reduction resources. Past efforts to assess societal vulnerability to lahar hazards have neglected to provide managers and policymakers with a characterization of population exposure to lahar hazards across multiple scenarios and among multiple jurisdictions.

The objective of this paper is to present a geospatial approach for characterizing and comparing community exposure to lahar hazards from multiple volcanoes. This process can help emergency managers, elected officials, and the general public to understand the scope of lahar issues across a region and where to potentially prioritize limited risk-reduction resources. To demonstrate this approach, we characterize and compare community variations in population exposure to lahar hazards for five active volcanoes in the State of Washington (USA). We integrate land cover, population, and economic datasets with mapped lahar-hazard zones from Mount Baker, Glacier Peak, Mount Rainier, Mount Adams and Mount St. Helens (Figure 1) to identify the number and distribution of people and businesses that occupy lahar-prone areas. We report on several aspects of population exposure to lahar hazards including (1) the distributions of populations within communities that occupy lahar-prone areas, (2) variations in overall community exposure, and (3) population growth trends in lahar-prone areas over time. Results of this study provide useful information to emergency managers, local officials, and the general public that will help them reduce risk and increase community resilience to volcano hazards.

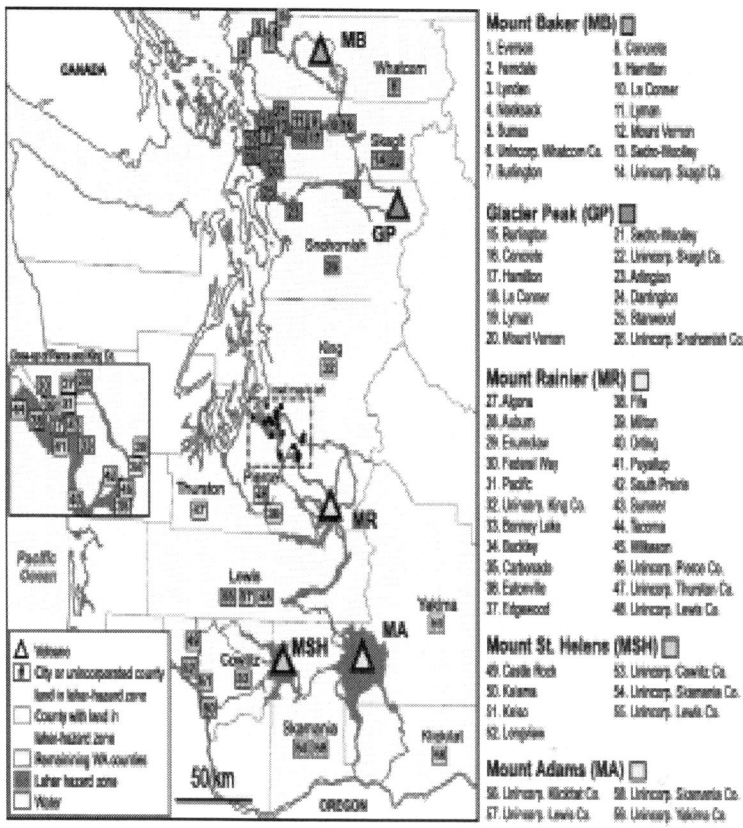

Figure 1: Map showing counties and incorporated cities within lahar-hazard zones associated with five active volcanoes in Washington.

Washington Volcanism

Washington State is home to five active stratovolcanoes—Mount Baker, Glacier Peak, Mount Rainier, Mount Adams, and Mount St. Helens—which form the northern-most section of the Cascade volcanic range in the United States (Figure 1). These volcanoes are the most active volcanoes in the Cascade Range, producing more than 90 eruptive episodes over the past 10,000 years (Siebert et al. [2010]). With the exception of Mount Adams, all have erupted in historical time, including the well-documented eruptions of Mount St. Helens in 1980–1986 and 2004–2008.

On average, lahars are the most common and farthest-reaching type of hazardous flow from eruptions at volcanoes in Washington State, outranking both pyroclastic flows and lava flows in rate of occurrence. These volcanoes collectively have produced more than 300 lahars from eruptive and noneruptive activity in the past 10,000 years (Hyde and Crandell [1978]; Beget [1982], [1983]; Frank [1983]; Pierson [1985]; Hoblitt et al. [1987]; Scott [1988]; Hildreth and Fierstein [1997]; Vallance [1999]) and they will likely produce more due to their steep slopes, extensive snow and ice cover, and hydrothermally altered summit areas. Geologic evidence suggests lahars from Mount Baker, Glacier Peak, and Mount Rainier have repeatedly reached the Puget Sound Lowland over the past 10,000 years, where today the majority of the state's population and economy are located. Past lahars from all Washington volcanoes have inundated valley bottoms more than 50 km away from their flanks and in some cases have exceeded 100 km (Hoblitt et al. [1987]).

Recognizing the threat that lahars pose to downstream communities, local, state, and federal government agencies have worked together on several efforts in Washington to prepare downstream communities and mitigate potential impacts. Volcano monitoring networks have been installed and maintained to detect precursory unrest as early as possible for timely hazard warnings (Guffanti et al. [2010]; Ewert et al. [2005]). Lahar hazard assessments have been completed that describe the types and extents of hazardous physical processes and to identify areas to focus risk-reduction efforts (Gardner et al. [1995]; Scott et al. [1995]; Wolfe and Pierson [1995]; Waitt et al.[1995]; Hoblitt et al. [1998]). Interagency coordination plans have been developed to support efficient emergency response by local, state, and federal agencies (Washington Military Department and Emergency Management [1999], Washington Military Department, Emergency Management Division [2012]). Educational products have been published and disseminated to promote awareness and preparedness (Mastin and Waitt [2000]; Scott et al. [2000]; Driedger and Scott[2008]; Dzurisin et al. [2008], [2013]).

Although much work has gone into overall volcano hazard awareness and monitoring both here in Washington and across the United States, less has been done to characterize societal vulnerability to these hazards, specifically the potential impacts to people (Aster et al. [2007]; National Research Council [2012]). There have been some attempts to examine resident perceptions of volcano threats, such as efforts at Mount St. Helens back in 1980 when the volcano first began to show activity (Perry et al. [1982]) and more recent explorations of risk perception from Mount Rainier hazards (Davis et al. [2006]; Johnston et al. [2006]). Wood and Soulard ([2009a], [b]) were the first studies to assess population exposure to lahar hazards in Washington but focused exclusively on Mount Rainier.

Study Area

This study of population exposure to lahar hazards in Washington focuses on 11 counties and 36 incorporated cities within them that intersect lahar-hazard zones associated with Mount Baker, Glacier Peak, Mount Rainier, Mount Adams, and Mount St. Helens (Figure 1). Three counties and seven communities are in lahar-hazard zones for multiple volcanoes. There are also 49 unincorporated towns and tribal reservations, as delineated by census-designated-places boundaries (U.S. Census Bureau [2010]) that intersect the lahar-hazard zones; however, because emergency services and land-use planning for these towns are performed by county offices, results related to the unincorporated areas within a county are reported at the county level.

The lahar-hazard zones shown in Figure 1 identify areas that could be affected by lahars generated in various river valleys that drain each volcano and are not meant to imply that all delineated areas would be affected by a single event. Each lahar-hazard zone is based on the extents of the maximum known or envisioned flows that have occurred at each volcano and therefore represent the most distal flow hazard zones. The known extent and distribution of lahar deposits from past events serve as a useful guide for establishing lahar-hazard boundaries, but a limitation exists when attempting to account for all possible and all likely scales of hazard extents. Therefore, it is important to note that these lahar-boundaries are not a prediction but serve merely as a guide to possible future extents. Within each of these distal lahar-hazard zones, hazard decreases with increasing distance down valley from each volcano as well as with increasing height above valley floors, and therefore not all areas within each zone are equally at risk. Additionally, the presence of several dammed reservoirs on rivers that drain some of these volcanoes (e.g., Baker Lake and Lake Shannon on Mount Baker; Alder Lake and Riffe Lake on Mount Rainier) may reduce the extent of lahar flow if they are drawn down in response to volcanic unrest and have enough storage to contain an eruption-generated lahar. Each lahar-hazard zone used in this study represents a different size event, recurrence interval, and degree of hazard and should be evaluated on an individual basis. Further descriptions of each lahar-hazard zone can be found in following volcano hazard assessments; Gardner et al. [1995] (Mount Baker), Waitt et al. [1995] (Glacier Peak), Hoblitt et al. [1998] (Mount Rainier), Scott et al. [1995] (Mount Adams), and Wolfe and Pierson [1995] (Mount St. Helens) along with the digital data series of mapped boundaries (Schilling [1996]; Schilling et al. [2008]).

Finally, this study focuses solely on lahar hazards and does not include other acute volcano hazards such as lava flows, pyroclastic density currents, or ballistic fall which primarily affect only areas proximal to volcanoes, where climate factors and land ownership (dominantly federal government) significantly limit the numbers of nearby residents. We also do not include ash fall hazards, which can strongly affect distal areas, but whose precise areas of impact can only be forecast on a day to day basis owing to variations in eruption style and magnitude and in wind directions.

METHODS

Community exposure to lahar-hazard zones at the five Washington volcanoes was assessed by using geographic information system (GIS) software to overlay geospatial data representing lahar-hazard zones, jurisdictional boundaries for incorporated cities and counties, and five indicators of human use (area of developed land and number of residents, employees, public venues, and dependent-care facilities). These variables represent data that U.S. jurisdictions are encouraged to collect as they develop State and local hazard mitigation plans (Federal Emergency Management Agency [2001]) to qualify for receiving certain types of hazard mitigation grant funds and other nonemergency disaster assistance under the U.S. Hazard Mitigation Grant Program in accordance with the Disaster Mitigation Act of 2000. National datasets were used to calculate all indicators to enable a systematic way of assessing community exposure over large areas with overlapping jurisdictions. We calculate the area of developed land and the number of individuals and businesses in lahar-prone areas to show emergency managers where hazard education may be most needed and where, in the absence of evacuations, potential losses could be greatest. We also calculate the percentage of these community assets that are in a lahar-hazard zone to provide insight about the relative impact of losses to an entire community. For example, if community A has 100 businesses in a lahar-hazard zone that represent 10 percent of the local economy and community B has 30 businesses in the lahar-hazard zone that represents 90 percent, then the relative impact of a future lahar may be greater in community B because it has a higher proportion of its businesses in the hazard zone. Economic recovery may be more challenging in community B, given the dramatic impact to businesses.

Developed land was identified using the 2006 National Land Cover Database (NLCD) (Fry et al.[2011]), which is a thematic land cover layer of the conterminous United States produced by automated classification routines of Landsat 7 Enhanced Thematic Mapper Plus (ETM+) and Landsat 5 Thematic Mapper (TM) imagery (30-m grid cells). We assume population exposure increases as the area and percentages of developed land within each lahar-hazard zone increases (Wood[2009]). We identified developed land using three 2006 NLCD classes: (1) high-intensity developed (>80% impervious surfaces, such as heavily built-up commercial and residential environments), (2) medium-intensity developed (50–79% impervious surfaces, such as single family housing and roads), and (3) low-intensity developed (20–49% impervious surfaces, such as single-family housing).

Various datasets were assembled to characterize the at-risk population. Resident counts were based on block-level population counts compiled for the 1990, 2000, and 2010 U.S. Census (U.S. Census Bureau [2010], [2012]). Census data assume a uniform distribution of population within a given block. If a census block was not entirely within a lahar-hazard zone, final population values were adjusted proportionately using the ratio of area within and outside the hazard zone. Employees and businesses were determined using the 2010 Infogroup Employer Database (Infogroup [2010]), which is a proprietary database that includes business locations, employee counts, and type based on the North American Industrial Classification System (NAICS; U.S. Census Bureau [2007]). We used NAICS codes to classify certain businesses as public venues (e.g., museums, overnight accommodations, and parks or other outdoor venues) and dependent-population facilities (e.g., child services, elderly services, medical centers, and K-12 schools). We highlight these two business types because individuals at dependent-care facilities may require evacuation assistance due to limited mobility (Wisner et al. [2004]), whereas individuals at public venues (e.g., tourists) may have limited situational awareness of volcano hazards. Our analysis serves as an approximation because we were unable to field verify the locations and attributes of the 8,807 businesses within the lahar-hazard zones of the five volcanoes; however, businesses with a physical address (i.e., not a Post Office box) were cross-referenced with the Homeland Security Infrastructure Program (HSIP) Gold 2011 database to improve the accuracy of our reporting. The amounts and percentages of all variables reported serve only as estimates of population exposure to lahar hazards in Washington State and do not take into account uncertainty in data sources, methods, and the spatial and temporal variability of each indicator.

Two composite indices were developed to compare community exposure to lahars for each of the geographic units that intersect lahar-prone areas for each volcano. Composite indices of the amount and percentage of each community's assets were developed by normalizing values in the five population indicators to the maximum value found within each category. Normalizing data to the maximum value creates a common data range of zero to one for each socioeconomic indicator and provides a simple approach for enabling comparisons among disparate data ranges. These values were added together for each geographic unit resulting in a composite score that ranged from zero to five. Each geographic unit has two composite indices that summarize the number and percentage of community assets in lahar-prone areas. To eliminate weighting bias between indices, a final score was then calculated for each geographic unit by normalizing each composite index to maximum values (range of zero to one for the two indices) and then adding the two indices resulting in a final combined score ranging from zero to two. These indices are unit-less, relative values with no absolute meaning for each community, but are used to compare the overall exposure of each community. Communities with higher scores have relatively higher numbers or percentages of population-related indicators and are therefore considered more vulnerable to future lahars.

RESULTS

Population Exposure among the Five Volcanoes

The lahar-hazard zones for the five active Washington volcanoes contain 274 km² of developed land, 191,555 residents, 108,719 employees, 433 public venues, and 354 dependent-care facilities. For each variable, Mount Rainier contains the highest percentage of regional (entire study area) totals in the lahar-hazard zones, ranging from 36% of the public venues to 54% of the employees (Figure 2). Mount Baker and Glacier Peak represent 21% to 30% of regional totals, respectively. Mount Adams and Mount St. Helens represent the lowest percentages at approximately 0% to 5% of regional totals, except for developed land at Mount St. Helens which is 12% of regional totals.

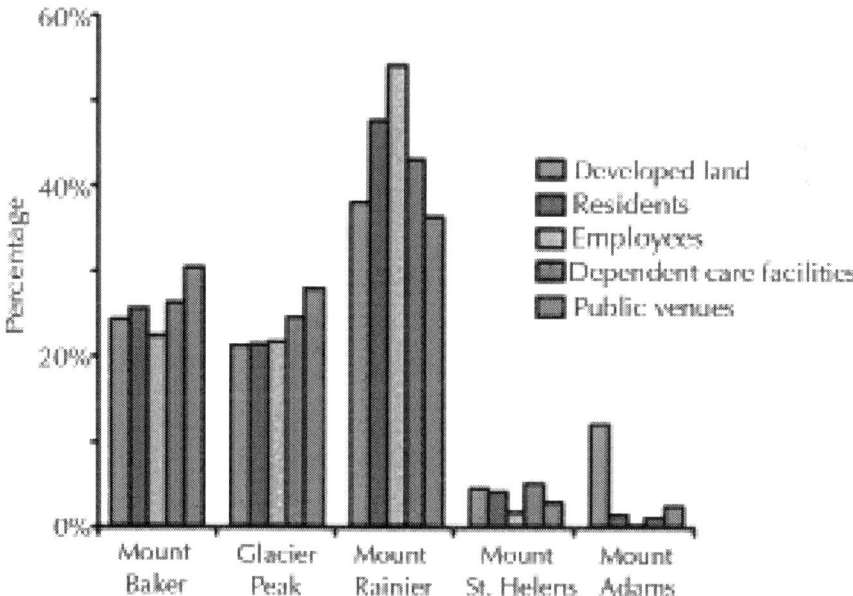

Figure 2: Percentage, by volcano, of the total amount of developed land and number of residents, employees, dependent-care facilities, and public venues in the lahar-hazard zones of the five active volcanoes in Washington.

Population Exposure at each of the Five Volcanoes

Mount Baker

The lahar-hazard zone of Mount Baker intersects eight incorporated cities and two counties (Figure 1) and contains 49,212 residents, 24,341 employees at 2,412 businesses, 132 public venues, and 93 dependent-care facilities. The majority of these variables are in the unincorporated areas of Skagit and Whatcom counties, followed by the incorporated cities of Burlington, Mount Vernon, and Sedro-Woolley (Figure 3). While these counties and cities have the highest numbers of populations and businesses in hazard zones, other communities have higher percentages of their people and businesses in the zones. For example, Mount Vernon has 5,426 residents in the lahar-hazard zone, but this represents only 17% of the community; whereas La Conner only has 891 residents in the zone, but they represent 100% of the community. Other small communities with high percentages of populations in the lahar-hazard zone include Nooksack, Sumas, and Hamilton. The one exception to this dichotomy of low numbers versus high percentages is the city of Burlington, which has both high numbers and high percentages. Of the 93 dependent-care facilities that are in the lahar-hazard zone, the majority are schools, adult residential care, child day-care centers, and outpatient-care facilities. The 193 public venues in the lahar-hazard zone include 51 religious venues, 30 overnight accommodations, and 26 parks. Certain public venues in the lahar-hazard zone are high-occupancy tourist sites, such as the Mount Baker Ski Area, the North Cascades National Park, and Mount Baker-Snoqualmie National Forest recreation areas.

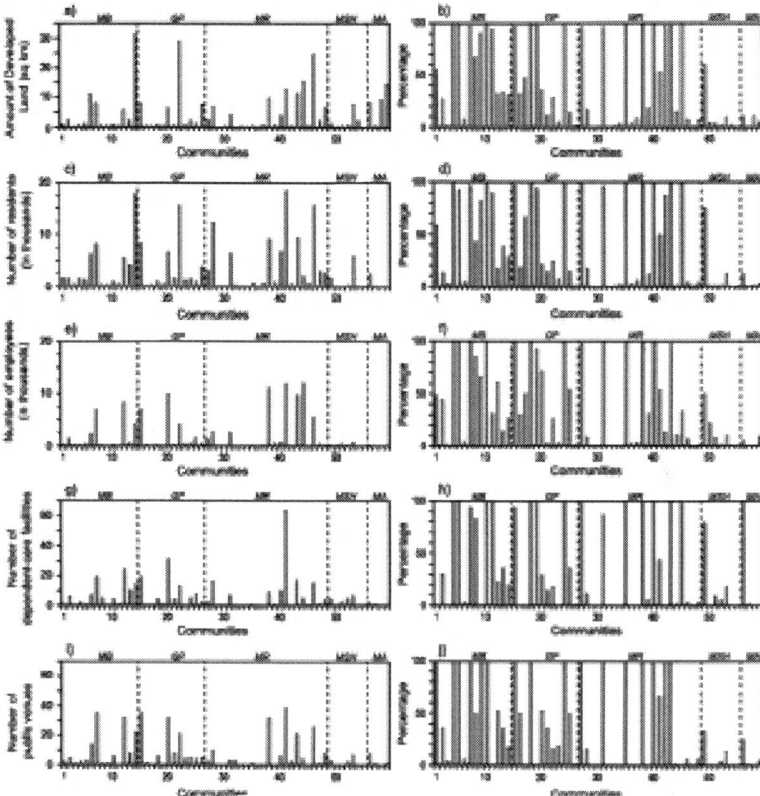

Figure 3: Counts and percentages of community assets in the lahar-hazard zones of the five active Washington volcanoes, including developed land (a and b), residents (c and d), employees (e and f), dependent-population facilities (g and h), and public venues (i and j). Communities are arranged along the x-axes geographically from north to south first by volcano and then by county, followed by an alphabetical listing of communities within each county. MB = Mount Baker, GP = Glacier Peak, MR = Mount Rainier, MSH = Mount St. Helens, MA = Mount Adams. Community numbering is listed in Figure 1.

Glacier Peak

The lahar-hazard zone of Glacier Peak intersects five incorporated cities and two counties (Figure 1) and contains 40,819 residents, 23,576 employees at 2,379 businesses, 121 public venues, and 87 dependent-care facilities. The section of the Glacier Peak lahar-hazard zone that follows the Skagit River significantly overlaps the lahar-hazard zone from Mount Baker, resulting in several communities (Burlington, Concrete, Hamilton, La Conner, Lyman, Mount Vernon, and Sedro-Woolley) exposed to lahars originating from both volcanoes. For Glacier Peak lahar-hazard zones, the highest amounts of the developed land, population, and business variables are in the unincorporated areas of Skagit and Snohomish counties and the incorporated cities of Burlington and Mount Vernon (Figure 3). Similar to Mount Baker, some communities contain high numbers of people and businesses in lahar-hazard zones (e.g., Mount Vernon) that represent low percentages of each variable while other communities contain low numbers but high percentages of their populations in hazard zones (e.g., La Conner, Darrington, and Lyman). As was the case with Mount Baker, Burlington is an exception with high numbers and high percentages in the Glacier Peak lahar-hazard zone. The majority of the 87 dependent-care facilities and 121 public venues in the lahar-hazard zone are in Burlington and Mount Vernon.

Mount Rainier

Among the five Washington volcanoes, the lahar-hazard zone of Mount Rainier (Figure 1) contains the highest number of incorporated cities (18), counties (4), residents (91,435), employees (58,969 at 3,821 businesses), public venues (158), and dependent-care facilities (153). Although there are 18 cities and 4 counties with land in the Mount Rainier lahar-hazard zone, the majority of the residents (86%), employees (75%), dependent-care facilities (90%) and public venues (87%) in the lahar-hazard zone are in cities of Puyallup, Auburn, Sumner, Fife, Orting, and Pacific, and the unincorporated portions of Pierce County (Figure 3). Tacoma is an exception to this list when considering exposed employees given the high number (12,224) and percentage of all employees in the zone (21%), which represent the concentration of employees near the Port of Tacoma. As was the case with Mount Baker and Glacier Peak, there are many smaller communities (e.g., Algona, Carbonado, South Prairie, and Wilkeson) with lower numbers of exposed populations that represent the majority, if not all in many cases, of the community. There are 151 dependent-care facilities (primarily K-12 schools) in the lahar-hazard zone, with most located in Puyallup. Schools (K-12 grade) are the most abundant and widely distributed type of dependent-care facility identified in the lahar-hazard zone. All of the K-12 schools in six communities (Pacific, Carbonado, Fife, Orting, Sumner, and Wilkeson) are in lahar-hazard zones, representing not only a life-safety issue but a long-term community recovery issue. There are 158 public venues in the lahar-hazard zone (primarily religious organizations, overnight accommodations, and parks), including high-occupancy sites such as casinos, the Puyallup Fairgrounds in Puyallup (over 1 million visitors each September; EventCorp Services [2011]), and Mount Rainier National Park (1.7 million visitors in 2010; National Park Service [2011]).

Mount St. Helens

The lahar-hazard zone of Mount St. Helens intersects four incorporated cities and three counties (Figure 1) and contains 7,645 residents, 1,656 employees at 151 businesses, 12 public venues, and 18 dependent-care facilities. The majority of each of these variables is in unincorporated Cowlitz County (Figure 3). Unlike the other volcanoes, the exposed land, populations, and businesses comprise a small percentage of community and county totals. The one exception is the city of Castle Rock, where in-hazard-zone percentages are high and range from 33% (public venues) to 80% (dependent-care facilities). Although community totals are relatively low for number of dependent-care facilities in the lahar-hazard zone, often these facilities represent each community's entire facility count, such as schools in Castle Rock and correctional facilities in Kelso. The public venues in the lahar-hazard zone include three parks, four religious venues, four overnight accommodations, and one library. Also in the lahar-hazard zone are access routes and recreational areas of the Mount St. Helens National Volcanic Monument, which draws more than 600,000 visitors to the Johnston Ridge Observatory each year (www.fs.usda.gov *webcite*; last visited 18 May 2013) and thousands of more visitors on the south and east sides of the monument.

Mount Adams

Of the five Washington volcanoes, the lahar-hazard zone of Mount Adams (Figure 1) contains the lowest number of residents (2,444), employees (177 at 44 businesses), public venues (10), and dependent-care facilities (3). The lahar-hazard zone crosses four counties (Klickitat, Lewis, Skamania, and Yakima) but no incorporated cities. The majority of the people and businesses are in unincorporated Klickitat County (Figure 3), but percentages are low in each of the four counties. The lahar-hazard zone contains three schools, six overnight accommodations, three religious venues, and one park.

Composite Indices of Community Exposure

Composite indices describing the amount and the percentage of assets in the lahar-hazard zone for each community and unincorporated area were developed using sums of normalized data in 5 categories—developed land, residents, employees, public venues, and dependent-care facilities. The City of Puyallup (Mount Rainier) has the highest composite amount index (4.4 out of 5), indicating that this community has the highest number of people and businesses in the lahar-hazard zone (Figure 4a). The only geographic area with a higher index in a category was unincorporated Skagit County for the amount of developed land it has in the Mount Baker lahar-hazard zone. Other communities and counties with high composite amounts include Skagit County (Glacier Peak); Pierce County, Sumner and Fife downstream of Mount Rainier; and Mount Vernon and Burlington for both Glacier Peak and Mount Baker lahar-hazard zones. Tacoma ranks eleventh behind these other communities and counties, due primarily to the highest number of employees in a lahar-hazard zone.

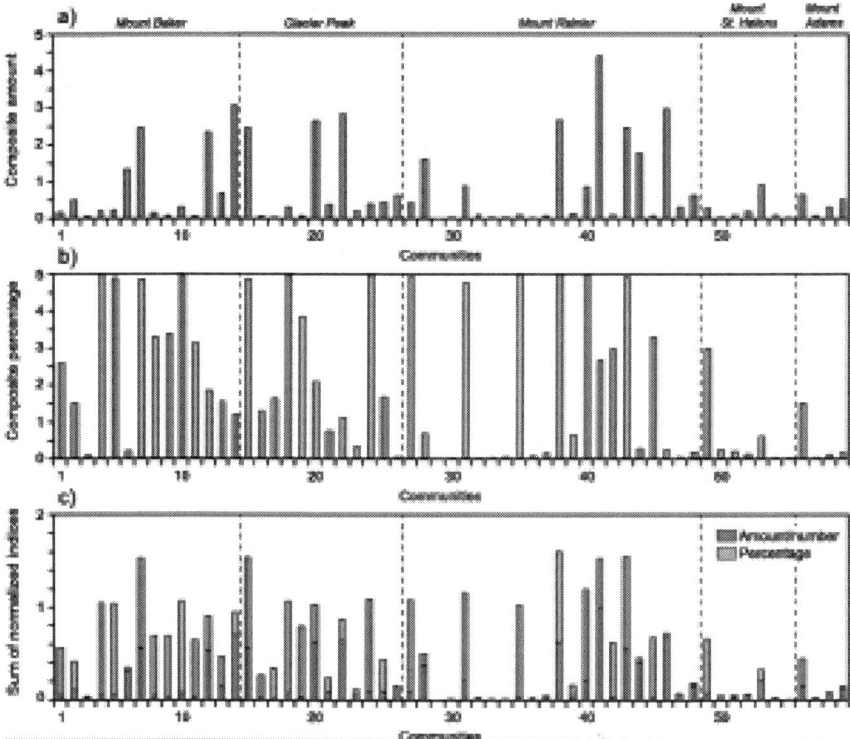

Figure 4: Composite indices summarizing (a) the amount of developed land and number of residents, employees, public venues, and dependent-care facilities in lahar-hazard zones, (b) the percentage of the community total of each of these variables in lahar-hazard zones, and (c) the normalized sum of the amount and percentage indices. Community numbering is listed in Figure 1.

Several communities have composite percentages of 4.8 to 5.0, indicating that they have the highest percentages of their developed land, populations, and businesses in the lahar-hazard zone for each of the five categories (Figure 4b). In all of these towns, the in-hazard populations and businesses represent close to or exactly 100% of the community. These types of communities include Darrington, La Conner, and Burlington downstream of Glacier Peak; Nooksack, La Conner, Sumas, and Burlington downstream of Mount Baker; and Fife, Orting, Carbonado, Sumner, Algona, and Pacific downstream of Mount Rainier.

A total of 15 communities have high (1.0 or greater) final composite scores, of which seven are downstream of Mount Rainier, four are downstream of Glacier Peak, and four are downstream of Mount Baker (Figure 4c). Certain communities are in this top tier of exposure due to the magnitude of exposed populations and businesses (e.g., Puyallup), whereas others have high percentages (e.g., Orting and Pacific). The two communities with the highest composite scores (Fife and Sumner) have moderately high amounts (and less than Puyallup) but some of the highest percentages of the various categories.

Clusters of communities with similar exposure to lahar hazards emerge on a plot of amount versus percentage indices for the five variables (Figure 5). The one exception is the city of Puyallup (41) downstream of Mount Rainier, which stands alone. Cluster A includes the city of Mount Vernon and unincorporated Skagit County (both for Glacier Peak and Mount Baker) and Pierce County (Mount Rainier) and represents areas with relatively high amounts but low percentages of the five variables. Cluster B communities (Sumner and Fife at Mount Rainier and Burlington at Glacier Peak and Mount Baker) have similar amounts as those in cluster A, but higher percentage index values (close to or equal to 1.0). Cluster C communities have similarly high percentage index values, but very low amount index values. Cluster D communities are similar to cluster C, but with lower percentage index values. Cluster E represents the bulk of the communities in the study area, with relatively low amount and percentage index values. Figure 5 also shows how the lower right of the graph (i.e., areas with amount index values greater than their percentage index values) is dominated by the unincorporated portions of the 11 counties.

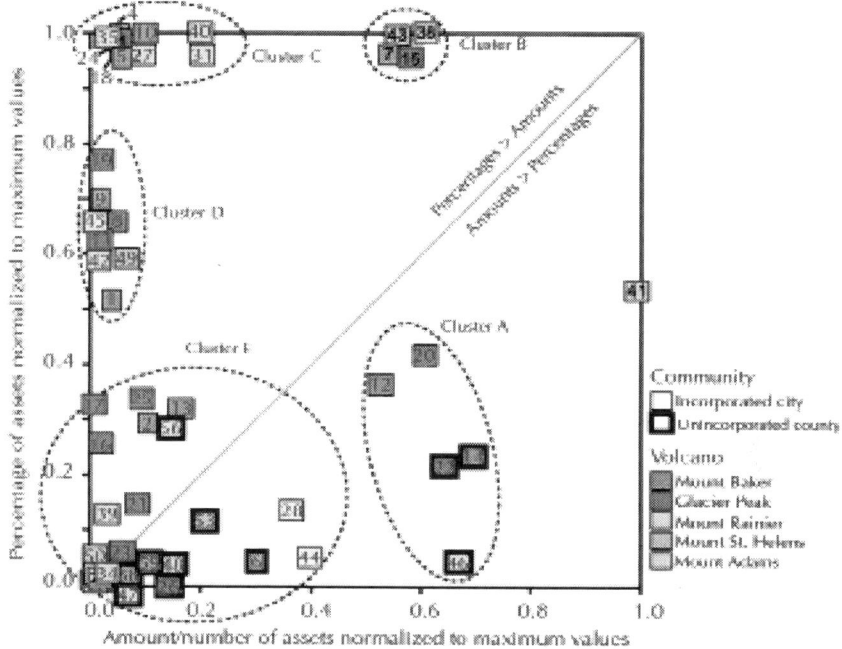

Figure 5: Comparison of normalized amount and percentage indices for communities in the lahar-hazard zones of the five active volcanoes in Washington. Unincorporated county land is identified with heavier outlines. Community numbering is listed in Figure 1. Any community not identified with its own colored square is located near the origin in cluster E and has values of less than 0.04 for both the relative amount and percentage of assets in lahar-hazard zones.

Changes in Community Exposure to Lahar Hazards between 1990 and 2010

Residential populations within lahar-prone areas of Washington State have increased by 48,080 residents over the 20-year time period between 1990 and 2010. More than half of this total increase is from population expansion within the Mount Rainier lahar-hazard zone (+24,619 residents), followed by smaller increases at Mount Baker (+12,003), Glacier Peak (+9,128), Mount St. Helens (+2,067), and Mount Adams (+263). At the community level (Figure 6a), the greatest increases in population during this time period were in the Mount

Rainier communities of Orting (+4,595 residents, representing a 214% increase) and Fife (+4,436 residents, representing a 94% increase), as well as Burlington for both Mount Baker and Glacier Peak (+3,900 residents, representing a 90–92% increase). The number of residents in lahar-hazard zones increased in all communities and counties except for small decreases in Enumclaw, Federal Way, Edgewood, Castle Rock, Kelso, Concrete, and the unincorporated areas of Lewis, Yakima, Snohomish, and King Counties.

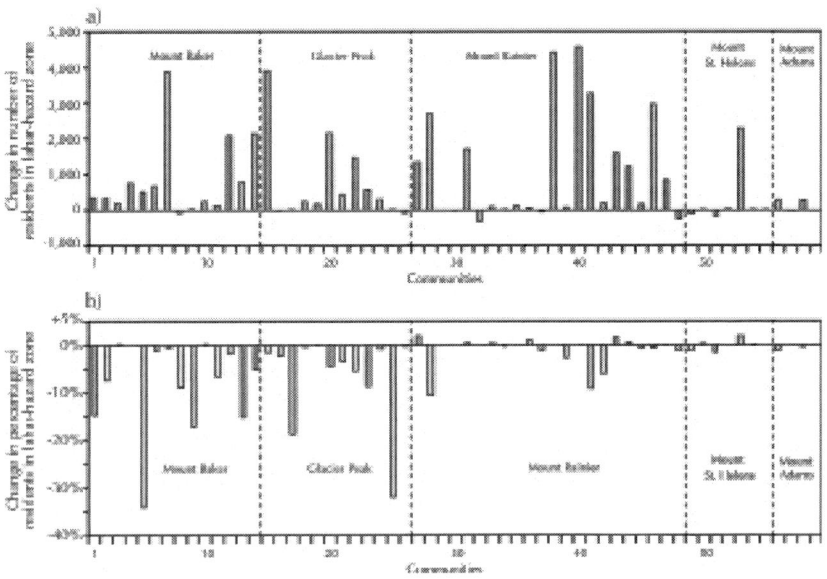

Figure 6: Community variations in residential population change between 1990 and 2010, in terms of (a) the change in the number of residents in a lahar-hazard zone and (b) the change in the percentage of a community's total resident population in a lahar-hazard zone. Community numbering is listed in Figure 1.

Although residential numbers increased through time, the percentage of residents in the lahar-hazard zone across the five Washington volcanoes (Figure 6b) remained the same between 1990 and 2010 (4.4% and 4.3%, respectively). Although Orting more than tripled and Fife, Algona, and Burlington nearly doubled their in-hazard-zone population, the communities saw little to no change in the overall percentage of their communities in the lahar-hazard zone. This is because each town was completely within lahar-hazard zones in 1990 and any new growth still resulted in an overall percentage of 100%. Decreases in the community percentage of exposed populations in most of the remaining communities and counties in the study area were not because the in-hazard-zone population decreased but because of much larger increases in residents elsewhere in the community. For the few towns that did have an increase in the community percentage of exposed populations, the increase was relatively small on the order of 1 to 2%.

DISCUSSION

Understanding how communities are vulnerable to lahars is a critical step for elected officials, emergency managers, and the public in their efforts to mitigate potential losses, as well as prepare for, respond to, and recover from future events. Previous efforts to characterize societal vulnerability to lahar hazards have focused largely on threats posed by single volcanoes. Missing from the literature are efforts to characterize and compare variations in community exposure to lahar hazards associated with multiple volcanoes. In the State of Washington, there are five active volcanoes, each having the potential to generate catastrophic lahars that could strike communities tens of kilometers downstream. In this section, we discuss the implications of our results on risk-reduction planning in Washington, as well as for lahar risk-reduction efforts in general throughout the world.

Results suggest that Washington lahar-hazard zones contain an estimated 191,555 residents, 108,719 employees at 8,807 businesses, 433 public venues that attract visitors, and 354 dependent-care facilities that house individuals who will need assistance to evacuate. Population exposure to lahar hazards varies greatly among eleven counties in Washington State. The communities with high numbers of assets tend to be larger in size (e.g., Puyallup, Mount Vernon, and Burlington) and represent high loss potentials. The communities with high percentages tend to be smaller in size (e.g., Nooksack, Orting), where loss of even a small number of assets may represent the entire community. Potential loss of a large number versus a high percentage of assets within communities represents information emergency managers, policymakers, and the general public can use in order to make decisions on where to focus risk reduction efforts if time and resources are limited (Wood and Soulard [2009a]).

Community clusters shown in Figure 5 can be used to identify communities with similar vulnerability issues. For example, the community and unincorporated areas of counties in cluster A may experience high losses from future events, but may be able to respond and recover with fewer external resources than other areas because the exposed populations and businesses represent low percentages of each community. Conversely, communities in clusters B and C may have a more difficult time recovering from a lahar given that all or the majority of their communities could be affected. Communities within similar clusters may wish to create partnerships to leverage limited risk-reduction resources. For example, although they are threatened by different volcanoes, the communities of cluster B (Sumner and Fife near Mount Rainier and Burlington near Mount Baker and Glacier Peak) have common issues of moderately high numbers of assets that constitute their entire communities. Similarly, a partnership might benefit the small cluster C communities (e.g., Orting, Pacific, Algona, and Carbonado near Mount Rainier and La Conner and Sumas near Mount Baker), which lie entirely in lahar-hazard zones. Risk-reduction successes may be transferred efficiently among communities in a single cluster.

When assessed in a state-wide perspective, Mount Rainier poses the biggest threat to assets in all sectors (Figure 2). The lahar-hazard zones of Mount Baker and Glacier Peak also contain high numbers of assets in each sector, the majority of each are from the same communities located in Skagit County where the lahar-hazard zones from each volcano significantly overlap in extent. The unincorporated areas of several counties are also exposed to multiple lahar-hazard zones (i.e., Glacier Peak and Mount Baker; Mount St. Helens and Mount Adams; Mount Rainier, Mount St. Helens and Mount Adams). Such cases present an educational outreach opportunity to raise awareness in communities about hazards related to multiple volcanoes.

The distribution and number of residential populations within each lahar-hazard zone may also warrant targeted education and preparedness efforts. The fact that significant portions of the residential population within the lahar-hazard zones of Mount Baker (49%), Glacier Peak (47%), Mount St. Helens (76%) and Mount Adams (100%)

reside in the unincorporated county areas (Figure 3) identifies a need for education outreach, awareness programs and evacuation planning for rural residents outside established community boundaries. In contrast, only 20% of residents in the lahar-hazard zone at Mount Rainier do not live in incorporated communities, indicating that community based awareness programs will likely reach a significant percent of the at-risk population.

Within each individual study area there are significant differences among communities in the types of population within lahar-prone areas (Figure 3). For example, within the Mount Baker lahar-hazard zone, the communities of Burlington and Mount Vernon and the unincorporated areas of Skagit County have high numbers of many types of people, whereas the exposed populations in other communities are dominated by certain types of populations, such as residents (e.g., Auburn and Orting) or employees (e.g., Tacoma and Stanwood). Education efforts will need to be tailored to each audience's needs in order to be most effective. For residents, sustained educational efforts that capitalize on existing social networks (e.g., city councils, neighborhood groups, schools, and parent and teacher associations) may be most effective (Wood and Soulard [2009a], [b]). Employees at businesses with large customer bases would ideally be trained in evacuation procedures; however, employees who reside in areas outside of lahar-hazard zones may be unaware of attendant hazards and will require access to educational materials (Wood and Soulard[2009a], [b]). Additional evacuation planning and staff training may be required in communities with high numbers of dependent-care facilities (e.g., Puyallup, Mount Vernon, and Burlington). The number and distribution of these various population groups provides essential information to officials when formulating risk-reduction strategies.

Residents and employees represent the dominant type of individuals within each hazard zone; however, because public venues and dependent-facilities do not include population counts at each facility it is difficult to compare actual population numbers between the five different indicator categories. The number and type of public venues and businesses in the lahar-prone areas of each community provides some insight about tourist populations but does not fully capture the range and magnitudes of tourist populations that may be exposed to lahar-hazards. An estimation of visitor statistics to areas near each volcano for recreation and tourism provides a first order assessment of transient populations within hazard prone areas but further research is warranted to better understand the number, distribution, demographics, and occupancy times to prepare and disseminate effective education and awareness information for these populations. Examples of targeted education outreach toward visitors and outdoor recreationists can be found at the visitor centers and on trailheads at Mount Rainier National Park and the Mount St. Helens National Volcanic Monument.

Examining changes in population exposure over time in lahar-prone areas provides insight on where new education efforts may be needed. Lahar education and evacuation training are long-term investments of time and resources and will not be one-time efforts. In Washington State, the number of people at risk from lahars continues to increase as residential populations, economic development, and recreation activities around these volcanoes increase. This is evident in some communities that have doubled (e.g., Burlington, Algona, and Fife) and tripled (Orting) their at-risk population in the last 20 years. Given the lack of lahar-hazard disclosure in property transactions, many of these new residents may be unaware of the lahar threat. In addition to increases in residential exposure, some of these communities (e.g., Burlington and Fife) also have relatively high numbers of public venues in their hazard zones, indicating the possibility of even more exposed populations.

Finally, although methods described in this paper provide insight on variations in population exposure to lahar hazards, they should not be construed as loss estimates. Results only summarize the spatial coincidence of populations and proxies for populations (e.g., public venues) with lahar-hazard zones. Unlike earthquakes that strike instantly and with little to no warning, lahars provide some level of warning due to instrumented warning systems, volcano monitoring networks that detect when a volcano is reawakening and moving toward eruption or hazard alerts spread through multiple channels of communication. With such warnings and a population aware of how to respond, people should be out of harm's way before a lahar reaches their communities. Beyond the immediate risk associated with lahar inundation, a significant problem for many communities will be managing displaced populations and dealing with long recovery time frames (years to decades) of damaged communities. To fully understand the threat that lahars pose to these communities, this exposure analysis will hopefully serve as a foundation for complementary studies designed to understand the perceptions, preparedness levels, likely behavior during an event, and general adaptive capacity of at-risk individuals. To date, there has been little work to gauge the perceptions and preparedness of at-risk individuals in this region. A series of surveys conducted in 2006 in Orting and Puyallup suggested that awareness of Mount Rainier lahar-hazards was high but that little had been done at the household level to prepare for future events (Davis et al. [2006]; Johnston et al. [2006]). Also lacking in this region are evacuation studies (e.g., Wood and Schmidtlein [2013]) to determine whether or not individuals would have sufficient time to successfully evacuate out of lahar-hazard zones.

CONCLUSIONS

This study provides a first-order assessment of population exposure to lahar-hazards in Washington State and is intended to further the dialogue on understanding societal risk in the region. Population-exposure analysis from multiple volcanoes and across multiple jurisdictions helps emergency managers understand and communicate where potential loss of life may be concentrated and where to focus risk-reduction efforts. Results presented here illustrate that although lahars are regional hazards that will impact communities in different ways, there are many commonalities in community exposure across multiple volcanoes. These results support the notion that a place-based context is important for understanding community vulnerability to volcano hazards (Jones and Andrey [2007]). A more regional comparative approach to assessing and addressing vulnerability to lahar-hazards is warranted, as opposed to one-size-fits-all mitigation and preparedness strategies that inadequately address differences in community context. Communities with common issues of vulnerability to lahar-hazards may wish to build partnerships to leverage limited resources, especially communities downstream of different volcanoes that may have not previously engaged in collaborative discussions. Data presented here will be used by emergency managers and local officials to help identify and tailor future preparedness, mitigation, recovery planning, and outreach activities within specific communities in Washington State. Although this study provides only a snapshot in time of community exposure to lahar-hazards in the State of Washington, its broad geographic coverage provides a regional scope that will allow emergency managers to identify hot-spot areas for more refined investigations related to adaptive capacity and resilience. In addition, as future data releases (e.g., decadal population counts) occur, additional research into how these systems change over time are warranted and could provide an on-going blueprint for risk-reduction planning across the region.

AUTHORS' CONTRIBUTIONS

AKD participated in the study's design, compiled various datasets, carried out data analyses, and drafted the manuscript. NJW conceived of the study, carried out analyses, produced figures, and drafted the manuscript. JWE participated in the study's design. All authors read and approved the final manuscript.

ACKNOWLEDGEMENTS

This study was a cooperative effort between the U.S. Geological Survey (USGS) Volcano Science Center and USGS Land Change Science Program. Special thanks to John Schelling (Washington State Military Department Emergency Management Division) and Mara Tongue (USGS) who provided support for this study and Jamie Ratliff (USGS) who provided the compiled 2010 Census data. We thank Jeff Peters and two anonymous reviewers for their helpful comments. Any use of trade names is for descriptive purposes only and does not imply endorsement by the U.S. Government.

REFERENCES

1. Aceves-Quesada JF, Diaz-Salgado J, Lopez-Blanco J (2007) Vulnerability assessment in a volcanic risk evaluation in Central Mexico through a multi-criteria-GIS approach. Nat Hazards 40:339-356

2. Aster R, Bergantz G, Carn S, Serafino G, Wilson J, Yepes H, White K (2007) Review of the United States Geological Survey Volcano Hazards Program. Report of the American Association for the

Advancement of Science Research Competitiveness Program, 35 p. [http://volcanoes.usgs.gov/publications/pdf/aaas2007.pdf]

3.   Auker MR, Sparks RSJ, Siebert L, Crosweller HS, Ewert J (2013) A statistical analysis of the global historical volcanic fatalities record. J Appl Volcanol 2(2):24

4.   Beget JE (1982) Postglacial volcanic deposits at Glacier Peak, Washington, and potential hazards from future eruptions.

5.   Beget JE (1983) Glacier Peak, Washington: a potentially hazardous Cascade volcano. Environ Geol 5:83-92

6.   Birkmann J (2006) Measuring vulnerability to promote disaster-resilience societies: conceptual framework and definitions. In: Birkmann J (ed) Measuring vulnerability to natural hazards: towards disaster resilience societies, United Nations University Press, Tokyo. pp 9-54

7.   Chester DK, Degg M, Duncan AM, Guest JE (2000) The increasing exposure of cities to the effects of volcanic eruptions: a global survey. Global Environmental Change Part B: Environ Haz 2(3):89-103

8.   Davis M, Johnston D, Becker J, Leonard G, Coomer M, Gregg C (2006) Risk perceptions and preparedness: Mt Rainier 2006 community assessment tabulated results. GNS Science Report 2006/17, Institute of Geological and Nuclear Sciences Limited.

9.   Driedger C, Scott W (2008) Mount Rainier – living safely with a volcano in your backyard.

10.   Dzurisin D, Stauffer PH, Hendley JW II (2008) Living with volcanic risk in the Cascades.

11.   Dzurisin D, Driedger CL, Faust LM (2013) Mount St. Helens, 1980 to now – what's going on? U.S. Geological Survey Fact Sheet. pp 2013–3014, (1.1) [http://pubs.usgs.gov/fs/2013/3014/]

12.   (2011) Washington State Fair: attendance and demographics.

13.   Ewert JW, Guffanti M, Murray TL (2005) An assessment of volcanic threat and monitoring capabilities in the United States: framework for a National Volcano Early Warning System NVEWS. U.S. Geological Survey Open-File Report 2005–1164

14.   Federal Emergency Management Agency (2001) State and local mitigation planning how-to-guide No. 2 – understanding your risks. FEMA 386–2 [http://www.fema.gov/library/viewRecord.do?id=1880]

15. Frank D (1983) Origin, distribution, and rapid removal of hydrothermally formed clay at Mount Baker, Washington.

16. Fry J, Xian G, Jin S, Dewitz J, Homer C, Yang L, Barnes C, Herold N, Wickham J (2011) Completion of the 2006 National land cover database for the conterminous United States. Photogramm Engin Rem S 77(9):858-864

17. Gardner CA, Scott KM, Miller CD, Myers B, Hildreth W, Pringle PT (1995) Potential volcanic hazards from future activity of Mount Baker, Washington.

18. Gregg CE, Houghton BF, Paton D, Swanson DA, Johnston DM (2004) Community preparedness for lava flows from Mauna Loa and Hualalai volcanoes, Kona, Hawaii. Bull Volcanol 66:531-540

19. Guffanti M, Diefenbach AK, Ewert JW, Ramsey DW, Cervelli PF, Schilling SP (2010) Volcano-monitoring instrumentation in the United States, 2008. U.S. Geological Survey Open-File Report 2009–1165

20. Hildreth W, Fierstein J (1997) Recent eruptions of Mount Adams, Washington Cascades, USA. Bull Volcanol 58:472-490

21. Hoblitt RP, Miller CD, Scott WE (1987) Volcanic hazards with regard to siting nuclear-power plants in the Pacific Northwest.

22. Hoblitt R, Walder J, Driedger C, Scott K, Pringle P, Vallance J (1998) Volcano hazards from Mount Rainier, Washington, revised 1998.

23. Hyde JH, Crandell DR (1978) Postglacial deposits at Mount Baker, Washington, and potential hazards from future eruptions.

24. InfoGroup (2010) Employer database: online dataset [http://referenceusagov.com/Static/Home, last accessed Oct 10, 2014]

25. Johannesdottir G, Gisladottir G (2010) People living under threat of volcanic hazard in southern Iceland: vulnerability and risk perception. Nat Hazard Earth Sys 10(2):407-420

26. Johnston D, Becker J, Coomer M, Ronan K, Davis M, Gregg C (2006) Children's Risk Perceptions and Preparedness: Mt Rainier 2006 Hazard Education Assessment Tabulated Results. GNS Science Report 2006/16, Institute of Geological and Nuclear Sciences Limited.

27. Jones B, Andrey J (2007) Vulnerability index construction: methodological choices and their influences on identifying vulnerable neighborhoods. Int J Emerg Manag 4(2):269-295

28. Kaye G, Cole J, King A, Johnston D (2009) Comparison of risk from pyroclastic density current hazards to critical infrastructure in Mammoth Lakes, California, USA, from a new Inyo craters rhyolite dike eruption versus a dacitic dome eruption on Mammoth Mountain. Nat Hazards 49(3):541-563

29. Kunzler M, Huggel C, Ramirez JM (2012) A risk analysis for floods and lahars: case study in the Cordillera Central of Colombia. Nat Hazards 64(1):767-796

30. Lavigne F, De Coster B, Juvin N, Flohic F, Gaillard J, Texier P, Mornin J, Sartohadi J (2008) People's behavior in the face of volcanic hazards: perspectives from Javanese communities, Indonesia. J Volcanol Geotherm Res 172:273-287

31. Mastin L, Waitt R (2000) Glacier Peak – history and hazards of a Cascade volcano.

32. National Park Service (2011) Mount Rainier annual visitor statistics 1967–2010. http://www.nps.gov/mora/parkmgmt/upload/vis-stats-1967-2010-2.pdf. Accessed 08 Jan 2011

33. (2012) Disaster Resilience: A National Imperative. The National Academies Press, Washington, DC.

34. Paton D, Millar M, Johnston D (2001) Community resilience to volcanic hazard consquences. Nat Hazards 24:157-169

35. Perry R, Lindell M, Greene M (1982) Threat perception and public response to volcano hazard. J Soc Psychol 116:199-204

36. Pierson TC (1985) Initiation and flow behavior of the 1980 Pine Creek and Muddy River lahars, Mount St. Helens, Washington. Bull Geol Soc Am 96:1056-1069

37. Pierson TC (1989) Hazardous hydrologic consquences of volcanic eruptions and goals for mitigative action: an overview. In: Starosolsky O, Melder OM (eds) Hydrology of disasters, Proc of the Technical Conference in Geneva, WMO, James and James, London (1989), pp. 220-236

38. Pierson TC, Wood N, Driedger C (2014) Reducing risk from lahar hazards—concepts, case studies, and roles for scientists. J Appl Volcanol 3(16):25

39. Rapicetta S, Zanon V (2009) GIS-based method for the environmental vulnerability assessment to volcanic ashfall at Etna Volcano. GeoInformatica 13(3):267-276

40. Schilling SP (1996) Digital Data Set of Volcano Hazards for Active Cascade Volcanoes, Washington.

41. Schilling S, Doelger S, Hoblitt R, Walder J, Driedger C, Scott K, Pringle P, Vallance J (2008) Digital data for volcano hazards from Mount Rainier, Washington. U.S. Geological Survey Open-File Report 2007- 1220

42. Scott KM (1988) Origins, Behavior, and Sedimentology of Lahars and Lahar-runout Flows in the Toutle-Cowlitz River System, Mount St. Helens, Washington.

43. Scott WE, Iverson RM, Vallance JW, Hildreth W (1995) Volcano Hazards in the Mount Adams region, Washington.

44. Scott KM, Hildreth W, Gardner CA (2000) Mount Baker – Living with an Active Volcano.

45. Scott KM, Luis Macias J, Naranjo JA, Rodriguez S, McGeehin JP (2001) Catastrophic debris flows transformed from landslides in volcanic terrains: mobility, hazard assessment, and mitigation strategies.

46. Scott KM, Vallance JW, Kerle N, Luis Macias J, Strauch W, Devoli G (2005) Catastrophic precipitation-triggered lahar at Casita volcano, Nicaragua: occurrence, bulking and transformation. Earth Surf Proc Land 30(1):59-79

47. Siebert L, Simkin T, Kimberly P (2010) Volcanoes of the World. University of California Press, Berkeley.

48. Spence RJS, Kelman I, Calogero E, Toyos G, Baxter PJ, Komorowski JC (2005) Modelling expected physical impacts and human casualties from explosive volcanic eruptions. Nat Hazards Earth Sys 5(6):1003-1015

49. Tayag JC, Punongbayan RS (1994) Volcanic disaster mitigation in the Philippines: experience from Mt. Pinatubo. Disasters 18(1):1-15

50. Tobin GA, Whiteford LM (2002) Community resilience and volcano hazards: the eruption of Tungurahua and evacuation of the Faldas in Ecuador. Disasters 25(1):28-48

51. U.S. Census Bureau (2007) North American Industry Classification System. U.S. Census Bureau. http://www.census.gov/epcd/www/naics.html. Accessed 01 April 2010

52. U.S. Census Bureau (2010) 2010 TIGER/Line® Shapefiles: U.S. Census Bureau. http://www.census.gov/cgi-bin/geo/shapefiles2010/main. Accessed 26 January 2012

53. U.S. Census Bureau (2012) American FactFinder: U.S. Census Bureau. http://factfinder2.census.gov/faces/nav/jsf/pages/index.xhtml. Accessed 03 June 2012

54. (1999) Postglacial Lahars and Potential Hazards in the White Salmon River System on the Southwest Flank of Mount Adams, Washington.

55. Voight B (1996) The management of volcano emergencies: Nevado del Ruiz. In: Scarpa R, Tilling RI (eds) Monitoring and Mitigation of Volcano Hazards, Springer, Berlin, Heidelberg. pp 719-769

56. Waitt RB, Mastin LG, Beget JE (1995) Volcanic-hazard Zonation for Glacier Peak Volcano, Washington.

57. (1999) Mount Rainier Volcanic Hazards Response Plan.

58. (2012) Mount Baker/Glacier Peak coordination plan: coordinating efforts between governmental agencies in the event of volcanic unrest at Mount Baker or Glacier Peak, Washington.

59. Wisner B, Blaikie P, Cannon T, Davis I (2004) At risk – Natural Hazards, People's Vulnerability and Disasters. Routledge, New York.

60. Witham C (2005) Volcanic disasters and incidents – a new database. J Volcanol Geotherm Res 148:191-233

61. Wolfe EW, Pierson TC (1995) Volcanic-Hazard Zonation for Mount St. Helens, Washington.

62. Wood N (2009) Tsunami exposure and estimation with land-cover data – Oregon and the Cascadia subduction zone. Appl Geog 29:158-170

63. Wood N (2011) Understanding risk and resilience to natural hazards. U.S. Geological Survey Fact Sheet 2011–3008 [http://pubs.usgs.gov/fs/2011/3008/]

64. Wood NJ, Schmidtlein M (2013) Community variations in population exposure to near-field tsunami hazards as a function of pedestrian travel time to safety. Nat Hazards 65(3):1603-1628

65. Wood NJ, Soulard CE (2009a) Community exposure to lahar hazards from Mount Rainier, Washington. U.S. Geological Survey Scientific Investigations Report 2009–5211

66. Wood NJ, Soulard CE (2009) Variations in population exposure and sensitivity to lahar hazards from Mount Rainier, Washington. J Volcanol Geotherm Res 188:367-378

Chapter

8

Reducing Risk from Lahar Hazards: Concepts, Case Studies, and Roles for Scientists

Thomas C Pierson[1], Nathan J Wood[2], and Carolyn L Driedger[1]

[1]U.S. Geological Survey, Volcano Science Center, Cascades Volcano Observatory, 1300 S.E. Cardinal Court, Suite 100, Vancouver 98683, WA, USA

[2]U.S. Geological Survey, Western Geographic Science Center, 2130 SW 5th Ave., Portland 97201, OR, USA

ABSTRACT

Lahars are rapid flows of mud–rock slurries that can occur without warning and catastrophically impact areas more than 100 km downstream of source volcanoes. Strategies to mitigate the potential for damage or loss from lahars fall into four basic categories: (1)

avoidance of lahar hazards through land-use planning; (2) modification of lahar hazards through engineered protection structures; (3) lahar warning systems to enable evacuations; and (4) effective response to and recovery from lahars when they do occur. Successful application of any of these strategies requires an accurate understanding and assessment of the hazard, an understanding of the applicability and limitations of the strategy, and thorough planning. The human and institutional components leading to successful application can be even more important: engagement of all stakeholders in hazard education and risk-reduction planning; good communication of hazard and risk information among scientists, emergency managers, elected officials, and the at-risk public during crisis and non-crisis periods; sustained response training; and adequate funding for risk-reduction efforts. This paper reviews a number of methods for lahar-hazard risk reduction, examines the limitations and tradeoffs, and provides real-world examples of their application in the U.S. Pacific Northwest and in other volcanic regions of the world. An overriding theme is that lahar-hazard risk reduction cannot be effectively accomplished without the active, impartial involvement of volcano scientists, who are willing to assume educational, interpretive, and advisory roles to work in partnership with elected officials, emergency managers, and vulnerable communities.

BACKGROUND

Lahars are discrete, rapid, gravity-driven flows of saturated, high-concentration mixtures containing water and solid particles of rock, ice, wood, and other debris that originate from volcanoes (Vallance [2000]). *Primary* lahars are triggered during eruptions by various eruption-related mechanisms; between AD 1600 and 2010 such lahars killed 37,451 people worldwide, including 23,080 in the 1985 Nevado del Ruiz disaster alone (Witham [2005]; Aucker et al. [2013]). During the same period *secondary* lahars, most commonly triggered by post-eruption erosion and entrainment of tephra during heavy rainfall, killed an additional 6,801 (Aucker et al. [2013]). Just in the past several decades, staggering losses from widely publicized lahar-related disasters at Mount St. Helens, USA; Nevado del Ruiz, Colombia; Mount Pinatubo, Philippines; and Mount Ruapehu, New Zealand, have demonstrated how lahars of both types significantly threaten the safety, economic well-being, and resources of communities downstream of volcanoes. Lahars can range in consistency from thick viscous slurries resembling wet concrete (termed *debris flows*) to more fluid slurries of mostly mud and sand that resemble motor oil in consistency (termed*hyperconcentrated flows*). These two types of flows commonly occur in all types of mountainous terrain throughout the world, but the largest and most far-reaching originate from volcanoes, where extraordinarily large volumes of both unstable rock debris and water can be mobilized (Vallance and Scott [1997]; Mothes et al. [1998]).

The destructive nature of lahars derives from their speed, reach, and composition—and our difficulty in predicting (in the absence of warning systems) when they may occur. Large lahars commonly achieve speeds in excess of 20 m/s on the lower flanks of volcanoes and can maintain velocities in excess of 10 m/s for more than 50 km from their source when confined to narrow canyons (Cummans [1981]; Pierson [1985]; Pierson et al. [1990]) (Table 1). Impact forces from multi-ton solid objects commonly suspended in debris-flow lahars (such as large boulders, logs, and other debris) and drag forces exerted by the viscous fluid phase can destroy almost any structure (Figure 1a). Hyperconcentrated-flow lahars damage structures primarily through vigorous lateral erosion of channels that results in bank collapse (Figure 1b). Both flow types commonly occur during a single lahar event as the highly concentrated head of a lahar typically transitions to a more dilute tail. On flow margins or at the downstream ends of depositional zones where velocities are much slower, lahars can encase buildings, roads, towers, and farm land in mud-rock slurries that can dry out to near concrete-like hardness. Yet fresh lahar deposits, commonly many meters deep, can remain fluidized like quicksand for days to weeks, complicating search and rescue efforts. Although most lahars are triggered during or shortly after volcanic eruptions, they can also be initiated without warning by noneruptive events, such as the gravitational collapse of structurally weakened volcanic edifices, large earthquakes, lake outbreaks, or extreme rainfall.

Table 1: Examples of lahar travel times from lahar source areas (points of initiation) to selected locations in downstream river valleys

Lahar date	Specified locations	Lahar trigger	Distance of specified location from lahar source (km)	Travel time from lahar source to specified location (min)	Average speed (meters per second)1
1926	Points along Hurano River, downstream of vent at Tokachidake volcano, Japan	Eruption (pyroclastic density current on snow and ice)	2.4	1	42
			6.4	4	26
			20.0	17	16
			23.2	29	5
1980	Points along Pine Creek, downstream of vent at Mount St. Helens, USA	Eruption (pyroclastic density current on snow and ice)	10.1	7	24
			14.1	11	16
			16.8	13	18
			20.9	19	14
1985	Point along Denjo River, downstream of landslide at Mount Ontake volcano, Japan	Earthquake-triggered slope failure	11.4	10	19
1985	Points along Río Chinchiná, downstream of vent at Nevado del Ruiz volcano, Colombia	Eruption (pyroclastic density current on snow and ice)	11.3	20	10
			33.0	70	7
			68.6	172	6

1990	Points along Drift River, downstream of vent at Redoubt Volcano, Alaska	Eruption (pyroclastic density current on snow and ice)	7.6	8	16
			18.5	27	10
1994	Points along Río Páez, downstream of vent at Nevado del Huila volcano, Colombia	Earthquake-triggered slope failures	4	2 to 3	22 - 33
			9	6 to 92	17 - 25
			30	20 to 302	17 - 25

Travel times are partly a function of lahar magnitude and are determined from eyewitness accounts, instrument recordings, and back-calculations of flow velocity based on physical evidence. Data from Pierson ([1998]) and Scott et al. ([2001]).

[1]Average speed is computed from the closest upstream point where timing information was available; not all timing points shown in this table. Localized lahar velocities tend to decrease as channel slopes decrease with distance downstream.

[2]Lahar could be heard or felt about 5 minutes before its arrival at this point.

Pierson *et al.*

Pierson *et al. Journal of Applied Volcanology* 2014 3:16, doi:10.1186/s13617-014-0016-4

Figure 1: Destructive effects of lahars. (a) Aerial view of Armero, Colombia, following destruction by a lahar on November 13, 1985, that killed approximately 21,000 people at this site alone (see Pierson et al.[1990]; USGS photo by R.J. Janda, 9 Dec 1985). Patterns of streets and building foundations are visible in the debris field at center of photo. **(b)** Aerial view of part of Angeles

City, downstream of Mount Pinatubo, Philippines, along the Abacan River, showing consequences of vigorous bank erosion by repeated post-eruption hyperconcentrated-flow lahars that were triggered by heavy monsoon rains (see Major et al. [1996]; USGS photo by TCP, 15 Aug 1991).

Various approaches to reduce and manage societal risks associated with lahar hazards have been applied over the years (Neumann van Padang [1960]; Smart [1981]; Suryo and Clarke [1985]; Pierson [1989]). These approaches fall into four basic categories of mitigation, including hazard avoidance, hazard modification, hazard warning, and hazard response and recovery (Figure 2). The goal of this paper is to provide an overview of each of these risk-reduction strategies and to highlight case studies of how (and how effectively) they have been applied at volcanoes around the world. The timing and magnitude of future lahars is uncertain and risk reduction efforts can be financially and politically costly; therefore economic, political, and social factors can compromise the implementation and long-term effectiveness of any strategy (Voight [1990], [1996]; Newhall and Punongbayan [1996]; Peterson [1996]; Prater and Lindell [2000]). We begin by discussing the importance of hazard and risk education for affected populations, elected officials, and emergency managers. We end by reemphasizing the call for committed involvement by volcano scientists in developing and executing these strategies. Scientist involvement improves the credibility and the efficacy of risk-reduction efforts. When the risks are perceived as credible and risk-reduction strategies are understood, tragic losses from future lahars on the scale of 20th-century lahar disasters can be avoided or at least minimized.

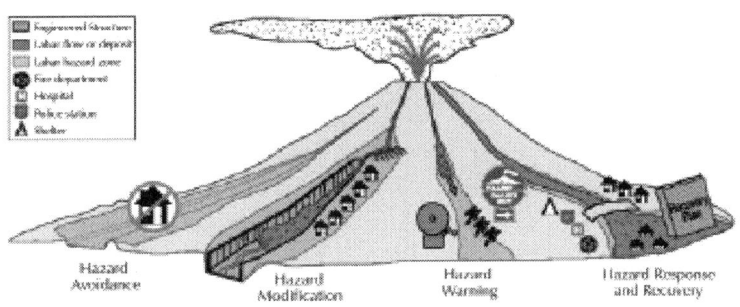

Figure 2: Schematic representation of the four basic strategies to reduce lahar-hazard risk within lahar hazard zones. Strategies include (1) hazard

avoidance with land-use planning and zonation; (2) hazard modification with engineered protection structures (bypass channel and deflection berm); (3) hazard warning to allow for timely evacuation; and (4) hazard response and recovery, which minimize long-term impacts after a lahar has occurred.

HAZARD AND RISK EDUCATION

The foundation for all risk-reduction strategies is a public that is well informed about the nature of hazards to their community, informed about how to lessen societal risk related to these hazards, and motivated to take risk-reducing actions. This knowledge base and accompanying appreciation of volcano hazards are needed to increase the interest and ability of public officials to implement risk-reduction measures and create a supportive and responsive at-risk population that will react appropriately when an extreme event occurs. Volcano scientists play a critical role in effective hazard education by informing officials and the public about realistic hazard probabilities and scenarios (including potential magnitude, timing, and impacts); by helping evaluate the effectiveness of proposed risk-reduction strategies; by helping promote acceptance of (and confidence in) hazards information through participatory engagement with officials and vulnerable communities as partners in risk reduction efforts; and by communicating with emergency managers during extreme events (Peterson [1988], [1996]; Cronin et al. [2004b]; McGuire et al.[2009]). But before successful use of hazard information can occur, the scientists' first and main role is to make technical data, hypotheses, and uncertainties understandable to non-technical users of hazard information. Serious misunderstandings can arise, sometimes with tragic consequences, when scientists do not perform this role effectively (Voight [1990]; Hall [1992]).

An effective hazard education program begins when scientists inform people in vulnerable communities about past hazardous events and current threats—information necessary for preparedness for future events. Scientists need to be involved in hazard-education efforts, because they provide the needed hazard expertise, and the public tends to imbue them with a high level of trust (Ronan et al. [2000]; Haynes et al. [2008]; Mei et al. [2013]). But the straightforward presentation of information that may seem logical to many scientists may not be effective; hazards information must be transmitted in ways that are not only understandable but also emotionally palatable and culturally relevant to the target audience (Cronin et al. [2004b]). People are more likely to implement risk-reduction strategies before an event or evacuate during an event if they comprehend that past events have impacted their communities, if they believe that future events could do so again and that viable mitigation options exist, and if they themselves have been involved in determining their community's risk-reduction strategies (Mileti [1999]). Community adoption of mitigation strategies is also more likely if hazard education is integrated into existing development programs and if it includes discussion of tangible actions that can be taken to protect lives and livelihoods, instead of just discussing uncontrollable threats (Paton et al. [2001]). The types of educational products, activities, and tasks that benefit from the active participation of scientists are varied (Figure 3):

Informative, jargon-free, general-interest publications and multi-media information products about potential hazards in digital and print formats (e.g., IAVCEI [1995], [1996]; USGS [1996], [1998],[2010]; Gardner et al. [2000]; Gardner and Guffanti [2006]; Driedger and Scott [2008]; Dzurisin et al. [2013]).

Technical information products to summarize scientific information about potential or ongoing volcanic activity or potential hazards, such as hazard-assessment reports, alerts and information statements on the status of current volcanic activity, volcanic-activity notification services, response plans developed in partnership with other agencies and stakeholders, and specific guidance based on the latest research (Guffanti et al. [2007]). Such products can be made available through print, fax, email, and social media outlets (e.g., Scott et al. [1997]; Hoblitt et al. [1998]; Pierce County [2008]; Wood and Soulard [2009a]).

Accessible and understandable spatial depictions of hazardous areas and evacuation routes to safe areas that are tailored to a target audience (Figure 3a,b), such as traditional hazard maps, evacuation route maps, explanations of the volcanic origins of familiar landscape features, labeled aerial photographs with vertical and oblique perspectives, and simple perspective maps keyed on cultural features and boundaries (Haynes et al. [2007]; Némath and Cronin [2009]). Web sites developed by local agencies can be good outlets for this type of information (e.g.,http://www.piercecountywa.org/activevolcano).

Hazards information presentations and training for the media (Figure 3c), emergency management officials (Figure 3d), first responders, land managers, public safety officials, search-and-rescue (SAR) teams, community-based monitoring teams, and public information officers before and during volcano crises (Driedger et al. [2008]; Frenzen and Matarrese [2008]; Peterson [1988],[1996]; Driedger et al. [2008]; Driedger and Scott [2010]; de Bélizal et al. [2013]; Stone et al.[2014]).

Teacher trainings (Figure 3e) and special school curricula for children in order to provide a foundation of knowledge at a young age, as well as to educate and motivate their families (e.g., Driedger et al. [2014]).

Presentations to and dialogues with community groups and councils, volunteer organizations, local government bodies, and schools about existing hazards (Figure 3f), while seeking opportunities to engage vulnerable populations in devising potential options for risk reduction (Peterson [1988],[1996]; Driedger et al. [1998]; Cronin et al. [2004a],[b]).

Relationship-building with communities and community leaders (official and unofficial) to establish trust and credibility, to encourage community-based risk-reduction solutions, and to maintain an ongoing dialogue with officials and at-risk community members (Peterson [1988], [1996]; Cronin et al. [2004b]; Haynes et al. [2008]; McGuire et al. [2009]; Mileti [1999]; Stone et al. [2014]).

Collaboration with emergency managers in the design and message content of signs for hazard awareness, locations of hazard zones, and evacuation procedures and routes (Figure 3g) (Schelling et al. [2014]; Driedger et al. [1998], [2002], [2010]; Myers and Driedger [2008a], [b]) and for disaster commemorations (such as monuments or memorials) that remind the public that extreme events are possible (Figure 3h).

Collaboration in the development of accurate and consistent warning messages to be sent out when a lahar triggers a warning system alert (Mileti and Sorenson [1990]).

Figure 3: Examples of some approaches for communicating hazards information to emergency managers, public officials, and at-risk populations. (a) Non-traditional hazard maps: An oblique perspective map showing potential lahar zones (brown) emanating from Mount Rainier volcano, with City of Tacoma, Washington (79 km downstream of Mount Rainier), in lower center of image along Puget Sound shoreline. Many people find it easier to visualize spatial information on such maps than on vertical plan-view maps. Satellite ground-surface image from Google Earth® modified by NJW, with Case 1

lahar hazard zones from Hoblitt et al. ([1998]) overlaid. (b)Signs and post-ers: A trail sign for hikers, using words and pictures, to convey lahar hazard information and instructions on what to do if they hear an approaching lahar (Mount Rainier National Park, USA). (c)Working with media: A USGS-hosted press conference to inform the media about the reawakening of Mount St. Helens (USA) in 2004 (USGS photo by D. Wieprecht). (d)Training: A training class on volcano hazards for emergency managers and given by scientists to provide an opportunity for relationship-building, as well as education (USGS photo by CLD). (e)Working with teachers: A scientist-led teacher workshop where simple physical models of lahars were used to help teachers grasp (and later teach) fundamental concepts about lahars (USGS photo by CLD). (f) Involving vulnerable populations in hazard-mitigation decisions: A 3-dimen-sional participatory mapping exercise for residents of a threatened village at Merapi volcano, Indonesia (photo by F. Lavigne, used with permission). (g) Practice drills: A lahar evacuation drill in 2002 at a school in Orting, Wash-ington, which is downstream of Mount Rainier (USGS photo by CLD).(h) Monuments and memorials: A simple disaster memorial commemorating 22 people killed by lahars in the town of Coñaripe on the lower flank of Villar-rica volcano, Chile, in 1964 (USGS photo by TCP).

Hazard education materials should be tailored to address the demographics and socioeconomic context of at-risk populations (e.g., Wood and Soulard [2009b]). This may include providing information in multiple languages on signs, pamphlets, and warning messages where appropriate, or conveying information in pictures or cartoons to reach children and nonliterate adults (Ronan and Johnston [2005]; Tobin and Whiteford [2002]; Dominey-Howes and Minos-Minopoulos [2004]; Gavilanes-Ruiz et al. [2009]). Educational outreach should also include efforts to reach tourists and tourism-related businesses, because these groups may lack hazard awareness and knowledge of evacuation procedures (Bird et al. [2010]).

A hazards and risk education program can increase its effectiveness by focusing outreach on those individuals and groups who can further spread information throughout a community. Such outreach can target institutions such as social organizations, service clubs, schools, and businesses, as well as trusted social networks (Paton et al. [2008], Haynes et al. [2008]). The key to sustaining hazard education is to identify and train community members with a vested interest in preparedness, such as emergency managers, educators, health advocates, park rangers, community and business leaders, and interested residents and other stakeholders. Training community members to integrate hazard information into existing social networks is especially crucial for hard-to-reach, potentially marginalized community groups, such as recent immigrants, daily workers coming from outside of hazard zones, or neighborhoods with people who don't speak the primary language (Cronin et al. [2004a]).

Direct involvement in training community members and elected officials extends a scientist's capacity to educate a community. It also provides opportunities for scientists to gain insight on how people conceptualize and perceive the hazards and the associated risks (for example, the role traditional knowledge and local experience), strengths and weaknesses of communication lines within a community, and any context-appropriate measures that might be used to increase local capacity for risk reduction (Cronin et al. [2004b]). Several studies have shown that people's behavior towards volcano risks is influenced not only by hazards information but also by the time since the last hazardous event and the interaction of their perceptions with religious beliefs, cultural biases, and socioeconomic constraints (Lane et al. [2003]; Gregg et al. [2004]; Chester[2005]; Lavigne et al. [2008]). Understanding these influences and the socio-cultural context of risk is important if scientists are to successfully change behaviors and not simply raise hazard awareness. Participatory methods such as three-dimensional mapping (Gaillard and Maceda[2009]) (Figure 3f), scenario planning (Hicks et al. [2014]), participatory rural appraisals (Cronin et al. [2004a][2004b]), and focus group discussions (Chenet et al. [2014]) can be used to understand the societal context of volcanic risk, to integrate local and technical knowledge, and to promote greater accessibility to information. These "bottom-up" efforts, as opposed to government-driven efforts that are perceived as "top-down", promote local ownership of the information (Cronin et al.[2004b]), empower at-risk individuals to implement change in their communities (Cronin et al.[2004a]), and can result in risk-reduction efforts becoming an accepted part of community thinking and daily life.

Finally, scientists should understand that effective hazard and risk education is a long-term investment of time and resources and will not be a one-time effort. One issue is that people may show great enthusiasm in hazards and risk information at public forums, but their interest and participation in risk-reduction activities may diminish over time as other day-to-day issues become higher priorities. Another issue is unavoidable turnover among users of hazards information. Elected officials may retire or be voted out of office. Emergency managers, first responders, and teachers may transfer to other positions or retire. People move in and out of vulnerable communities. So, just as scientists continually monitor changing physical conditions at volcanoes, they should also appreciate the dynamic nature of the perceptions and knowledge of hazards within communities, agencies, and bureaucracies—and plan for sustained education and outreach efforts.

STRATEGIES FOR LAHAR-HAZARD RISK REDUCTION

Each of the four basic risk-reduction strategies of hazard avoidance, hazard modification, hazard warning, and hazard response and recovery (Figure 2) has basic underlying requirements for successful application. These requirements include an accurate assessment of the hazard; a realistic understanding by elected officials, emergency managers, and at-risk populations of the hazards, risks, and limitations of any implemented strategy; thorough planning; adequate funding; practice exercises and drills, where appropriate; and effective communication among stakeholders during actual lahar occurrence (Mileti [1999]; Leonard et al. [2008]). Scientists have important roles to play in all of these underlying requirements.

Hazard Avoidance

A range of approaches can either regulate or encourage hazard avoidance—the strategy seeking to expose as few lives and societal assets as possible to potential loss. Land-use zoning regulations or development of parks and preserves that ban or limit occupation of hazard zones are ways to keep people, developed property, and infrastructure out of harm's way. Another way is for local government policies to allow occupation of hazard zones but to also impose disincentives for those who choose to live there. A third way is to educate the public about the hazard, the risks, and the probabilities of hazardous event occurrence, and then to trust that people will choose to minimize the hazard exposure of their homes and businesses.

A complete ban on development in a hazard zone is probably the most effective way to avoid the hazard. This may be easiest immediately following a disaster and if the ban aligns with cultural values, such as when the entire town site of Armero, Colombia, was made into a cemetary after about 21,000 people were killed there by a lahar in 1985 (Pierson et al. [1990]; Voight [1990]). However, it is commonly challenging to implement development bans based on hazard zonation prior to a disaster due to people's strong attachment to a place, cultural beliefs, political push-back from business and real-estate interests, the lack of alternative locations for new development, attitudes of individuals who don't want to be told where they can or cannot live, or needed access to livelihoods that exist in volcano hazard zones (Prater and Lindell [2000]; Lavigne et al. [2008]). Indeed, lahar hazard zones can be attractive for transportation and other infrastructure and for residential development, because these areas typically encompass deposits of previous lahars that offer flat topography, commonly above flood hazard zones, and they may offer scenic views of a nearby volcano (Figure 4). Lahar and related deposits also may be attractive for resource extraction. In the Gendol valley at Mount Merapi (Indonesia) for example, thousands of people work daily as miners in high-hazard zones, excavating sand and gravel to sell. Most, if not all, are aware of the risk but are willing to accept it because of the financial reward (de Bélizal et al.[2013]). In other cases such hazard zones may already be occupied by well-established communities—a reality that makes development bans problematic. A strong cultural attachment to the land and the lack of available safe land elsewhere may lead communities to accept lahar risks and even continue to rebuild homes after multiple lahar burials (Crittenden [2001]; Crittenden and Rodolfo [2002]).

Figure 4: Mount Rainier volcano and dense residential housing in down-stream community of Orting, Washington. The town is built on the flat up-per surface of a lahar deposit from Mount Rainier that was emplaced about 500 years ago. Orting is one of several communities that are in lahar hazard zones downstream of Mount Rainier. A warning system in this valley would give residents about 40 minutes to evacuate to high ground (USGS [2013]). USGS photograph by E. Ruttledge, 18 Jan 2014.

A more realistic land-use planning approach may be to restrict the kind or amount of development allowed to occur in lahar hazard zones. For example, vulnerable valley floors could be limited to agricultural use only, with homes built on higher ground. Downstream of Mount Rainier in Pierce County (Washington, USA), comprehensive land use plans include urban growth boundaries that prohibit tourist facilities larger than a certain size and limit other high-density land uses in lahar hazard zones (Pierce County [2014]). Downstream of Soufriére Hills volcano in Montserrat (British West Indies), only daylight entry into certain hazard zones for farming was allowed in the 1990s, due to pyroclastic-flow and lahar hazards associated with the actively erupting volcano (Loughlin et al. [2002]). The goal of such restrictions is to minimize population exposure and to only allow land uses in which people could be evacuated quickly, yet such measures are not always foolproof (Loughlin et al. [2002]). Ordinances can also limit the placement of critical facilities (hospitals, police stations, schools, and fire stations) in hazard zones, so that basic community services would be available for rescue, relief, sheltering, and recovery efforts in the event of a lahar (Pierce County [2014]).

Where no restrictions are imposed on development of lahar hazard zones, it may be possible to discourage development through the use of various disincentives. These could include higher property tax rates, higher insurance rates, and limitation of public services or infrastructure in designated hazard zones. For example in the United States, the National Flood Insurance Program requires that people living in designated flood zones purchase flood insurance (Michel-Kerjan[2010]). As premiums for such types of insurance increase, purchase of a home in a hazard zone should become less attractive.

Hazard education alone could, theoretically, also achieve some hazard avoidance, but evidence suggests that many residents already living in hazard-prone areas rarely undertake voluntary loss-prevention measures to protect their property, despite increased hazard awareness (Michel-Kerjan [2010]). Discouraging new residents from moving into hazard zones may be more realistic. Focused public education campaigns are one way to raise hazard awareness. Another is to require that hazard information be disclosed to people buying property or building structures in a hazard zone. Such disclosures are required on building-permit applications in Orting, Washington in the lahar hazard zone downstream of Mount Rainier. Some individuals may use increased hazard awareness to assess whether the risk is acceptable, others may not, and still other may object to increased hazard awareness. In fact, just the dissemination of hazards information to people living in hazard zones can engender fierce political opposition, particularly from some business and real-estate interests (Prater and Lindell [2000]).

Volcano scientists play important supporting roles throughout any land-use planning process aimed at reducing risk from lahar hazards. First, land-use decisions require hazard-zonation maps that are scientifically defensible, accurate, and understandable, given the potential for political, social, or legal push-back from various constituents. Second, good planning needs input from predictive models that estimate lahar runout distances, inundation areas, and travel times to populated areas. In addition, scientists are needed to help explain the uncertainties inherent in the maps and models, to estimate the likelihood of occurrence, and to evaluate the effectiveness of proposed risk-reduction strategies as land-use planners balance public safety against economic pressures to develop.

Hazard Modification

Some communities predate recognition that they are situated in a lahar hazard zone. Others may expand or be developed in hazard zones because of social and economic pressures, inadequate understanding of the risks, or acceptance and tolerance of the risks. When societal assets are already in lahar hazard zones, construction of engineered protection structures can reduce risk by (a) preventing some lahars from occurring, (b) weakening the force or reach of lahars, (c) blocking or trapping lahars before they can reach critical areas, or (d) diverting lahars away from critical areas—all methods of hazard modification (Smart [1981]; Baldwin et al. [1987]; Hungr et al. [1987]; Chanson [2004]; Huebl and Fiebiger [2005]). Engineered protection works, sometimes referred to as sabo works (sabō = "sand protection" in Japanese), and slope stabilization engineering methods have been widely used for centuries in volcanic areas in Japan and Indonesia, as well as in the Alps in Europe for protection from nonvolcanic debris flows.

Engineered structures designed for lahar protection downstream of volcanoes have many of the same advantages and disadvantages of river levees in flood-prone areas, sea walls in coastal areas, or engineered retrofits to buildings and bridges in seismic areas. The main advantages of this approach are that communities can survive small- to moderate-size events with little economic impact, and communities, if they choose to, can gradually relocate assets out of hazard zones. However, protection structures are expensive to build and maintain, which may overly burden communities financially or lead to increased vulnerability if funding priorities shift and maintenance is neglected. Another important disadvantage is that protection structures tend to lull populations into a false sense of security. People commonly assume that all risk has been eliminated, and this perception may result in fewer individuals taking precautionary steps to prepare for future events. This view may also result in increased development of areas now perceived to be safe because of the protective structure. The reality is that risk is eliminated or reduced only for events smaller than the 'design event' that served as the basis for construction. Events larger than the design event can occur and when they do, losses can be even larger because of the increased development that occurred after construction of the protection structure—also referred to as the 'levee effect' in floodplain management (Tobin [1995]; Pielke [1999]). This was the case near Mayon Volcano (Philippines) where lahar dikes built in the 1980s led to increased development behind the structures. When they failed because of overtopping by lahars during Typhoon Reming in 2006, approximately 1,266 people were killed (Paguican et al. [2009]). The effectiveness and integrity of engineered structures can also be compromised by the selection of cheap but inappropriate construction materials (Paguican et al. [2009]) and by ill-informed human activities, such as illegal sand mining at the foot of structures or dikes occasionally being opened to allow for easier road access into communities. Therefore, although protection structures may reduce the number of damaging events, losses may be greater for the less frequent events that overwhelm the structures. In addition, engineered channels and some other structures can have negative ecological effects on watersheds.

The potential for large losses is exacerbated if public officials choose to build the structure that is affordable, rather than the structure a community may need. Economics and politics may play a bigger role than science in deciding the type, size, and location of protection structures, because of the high financial costs and land-use decisions associated with building the structures and with relocating populations that occupy construction areas (Tayag and Punongbayan [1994]; Rodolfo[1995]) (Case study 1). Because decision makers will have to balance risk against cost, scientists have a significant role in helping public officials by (a) estimating the maximum probable lahar (the design event); (b) predicting probable flow routes, inundation areas, and possible composition and flow-velocity ranges; (c) estimating probabilities of occurrence; and (d) evaluating the effectiveness of proposed mitigation plans and structures.

Case Study 1: when Economics and Politics Trump Science

Following the June 15, 1991, eruption of Mount Pinatubo (Philippines), lahars and volcanic fluvial sedimentation threatened many downstream communities. Geologists from a number of institutions met with officials at local, provincial, and national levels to explain the threats and to evaluate and discuss proposed countermeasures. Due to political pressures (Rodolfo[1995]), officials ultimately adopted a lahar mitigation strategy that was based on the construction of parallel containment dikes close to the existing river channels, using easily erodible fresh sand and gravel deposits of earlier lahars as the construction material. Appropriation of the private land needed for lahar containment areas of adequate size was viewed by officials as too politically costly. Officials hoped the dikes would divert lahars and floods past vulnerable communities. However, nearly all the geologists involved in the discussions expressed the opinion that this was a poor strategy because (a) channel gradients were too low for effective sediment conveyance and deposition would occur in the wrong places, (b) dike placement did not provide adequate storage capacity and dikes would be overtopped or breached, (c) most of the dikes were not revetted and would be easily eroded by future lahars, and (d) people would be lured back to live in still-dangerous hazard zones. The advice of the scientists was not heeded, and over the next several years many of these predictions came true, including breached dikes due to lahar erosion and overtopped dikes due to sediment infill. Lahars breaking through the levees caused fatalities and destroyed many homes. A government official later explained (to TCP) that political considerations prompted the decisions to minimize the area of condemned land and build lahar catch basins that were too small. He felt that the plan recommended by the geologists would have angered too many people and that it was better for officials to be seen doing something rather than nothing, even if the chance of success was low. Indeed, political and economic forces can override scientific recommendations (Tayag and Punongbayan[1994]; Rodolfo[1995]; Janda et al.[1996]; Newhall and Punongbayan[1996]; Crittenden[2001]).

Slope Stabilization and Erosion Control

Volcanic ash mantling hillslopes is extremely vulnerable to rapid surface erosion and shallow landsliding, and it is easily mobilized as lahars by heavy rain (e.g., Collins and Dunne [1986]; Pierson et al. [2013]). Even after long periods of consolidation and revegetation, ash-covered slopes can fail on massive scales and result in catastrophic lahars (Scott et al. [2001]; Guadagno and Revellino [2005]). Various methods of slope stabilization, slope protection, and erosion control can limit shallow landsliding or surface erosion in disturbed landscapes that could produce extreme sediment inputs to rivers (Figure 5), although most of these approaches are intensive, costly, and generally limited to hillside-scale problem areas (see overviews in Theissen [1992]; Morgan and Rickson [1995]; Gray and Sotir [1996]; Holtz and Schuster [1996]; Schiechtl and Stern [1996]; Beyers [2004]; Valentin et al. [2005]). These are only briefly summarized here. Options for drainage-basin-scale slope stabilization and erosion control are more limited, have been tested mostly in basins disturbed by wildfire rather than by volcanic eruptions, and are not always effective (Beyers [2004]; deWolfe et al. [2008]).

Figure 5: Example of slope stabilization. Timber retaining walls used to stabilize a steep slope in a volcanic area in Japan (USGS photo by TCP).

Regardless of scale of application, slope stabilization and erosion control techniques attempt to either (a) prevent shallow landsliding by mechanically increasing the internal or external forces resisting downslope movement, decreasing the forces tending to drive downslope movement, or both; or (b) prevent rapid surface erosion and sediment mobilization on slope surfaces and in rills, gullies, and stream channels (Gray and Sotir [1996]; Holtz and Schuster [1996]). Inert materials used to stabilize slopes and control erosion include steel, reinforced concrete (pre-cast elements or poured-in-place), masonry, rock, synthetic polymers, and wood, although many of these degrade and weaken with time. Biotechnical stabilization (Morgan and Rickson [1995]; Gray and Sotir[1996]) uses live vegetation to enhance and extend the effectiveness of many engineered structures.

Forces resisting slope failure or erosion can be maintained or augmented by a variety of approaches (Morgan and Rickson [1995]; Gray and Sotir [1996]; Holtz and Schuster [1996]). Counterweight fills, toe berms, retaining walls, and reinforced earth structures can buttress toes of slopes. To maintain buttressing at a toe slope, revetments using riprap, gabion mattresses, concrete facings, and articulated block systems can prevent toe-slope erosion. Anchors, geogrids (typically wire-mesh mats buried at vertical intervals in a slope face), cellular confinement systems consisting of backfilled three-dimensional structural frameworks; micro-piles, deeply rooted woody vegetation, chemical soil binders, and drains to decrease internal pore pressures can increase the shear strength of natural or artificial slopes. To reduce the driving forces, proven methods include regrading to lower slope angles, and weight reduction of structures or materials placed on slopes. Surface erosion of slopes can be controlled by protecting bare soil surfaces and by slowing or diverting surface runoff through the application of reinforced turf mats, geotextile and mulch blankets, hydro-seeded grass cover, and surface drains. Channelized surface erosion can be retarded with gully fills or plugs of cut brush or rock debris, or small check dams.

Intensive slope-stabilization and erosion-control techniques such as many of those listed above may be too costly for large areas of volcanically disturbed drainage basins, but they may be cost-effective in specific problem areas. Over large areas, economically feasible approaches may include tree planting, grass seeding, and grazing management to limit further destruction of slope-stabilizing vegetation. However, much post-disturbance erosion is likely to occur before grass seed can germinate or tree seedlings can grow to effective size, and a number of studies have shown that large-scale aerial grass seeding is no more effective for erosion control than the regrowth of natural vegetation (deWolfe et al. [2008]).

Lake Stabilization or Drainage

Stabilizing or draining lakes that could breach catastrophically without warning is another way to prevent lahars from reaching vulnerable downstream areas. Crater lakes, debris-dammed lakes (dammed by pyroclastic-flow, debris-avalanche, or lahar deposits), and glacial moraine-dammed lakes all can become unstable if their impounding natural dams are overtopped or structurally fail. Historic rapid lake outbreaks in several countries have triggered catastrophic lahars that resulted in loss of life (O'Shea [1954]; Neumann van Padang [1960]; Umbal and Rodolfo [1996]; Manville[2004]). Very large prehistoric outbreaks of a volcanically dammed lake have been documented having peak flows comparable to the world's largest floods (Scott [1988]; Manville et al. [1999]). Stabilization methods include armoring of existing spillways on natural dams, construction of engineered spillways, and rerouting lake outflow by pumping or drainage through tunnels (Sager and Chambers [1986]; Willingham [2005]) (Figure 6; Case study 2). Preemptive drainage of dangerous lakes can be fraught with difficulties and may not be successful (Lagmay et al. [2007]).

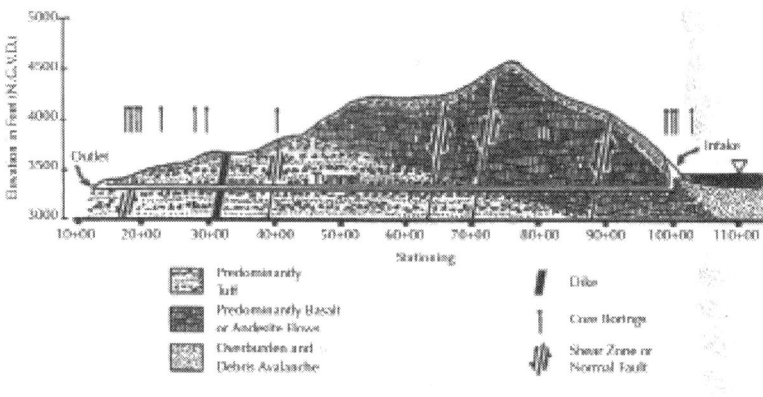

Figure 6: Lake-level stabilization to prevent failure of a natural debris dam and a subsequent lahar. At Mount St. Helens (USA) a tunnel was bored through a mountain ridge to divert water from Spirit Lake into an adjacent drainage basin. In this case debris-avalanche and pyroclastic-flow deposits formed the potentially unstable natural dam. This geologic cross section shows the 2.5-km-long outlet tunnel, which stabilizes the lake by keeping the water surface at a safe level below the dam crest (from Sager and Budai [1989]).

Case Study 2: Examples of Lake Stabilization

Since AD 1000, 27 eruptions of Mount Kelud (Java, Indonesia) have catastrophically expelled lake water from the volcano's crater lake and created several deadly lahars, including a lahar in 1919 that killed more than 5000 people (Neumann van Padang[1960]). In an attempt to drain this lake, engineers in 1920 dug a drain tunnel over 955 m in length from the outer flank of the cone into the crater but eventually abandoned the project because of ongoing volcanic activity and other technical difficulties. Thereafter, siphons were constructed to control the lake level, and these were responsible for partial drainage of the crater lake and for a reduced number of lahars during the 1951 eruption (Neumann van Padang[1960]).

More recently, debris-avalanche and pyroclastic-flow deposits from the 1980 eruption of Mount St. Helens (Washington, USA) blocked tributary drainages of the North Fork Toutle River and enlarged several preexisting lakes. The largest and potentially most dangerous of these was Spirit Lake, which, when mitigation efforts began, was impounding 339 million m3of water—enough to form a lahar that could have destroyed major parts of several cities located approximately 90 km downstream. To prevent the Spirit Lake blockage from ever being breached by overflow, the level of the lake surface was stabilized by the U.S. Army Corps of Engineers (USACE) at a safe level, first by pumping water over the potentially unstable natural dam in pipes using diesel pumps mounted on barges, and thereafter by draining lake water through a 3.3-m-diameter outlet tunnel that was bored 2.5 km through an adjacent bedrock ridge to form a permanent gravity drain that was completed in 1985 (Figure6). The USACE stabilized the outlets from two other debris-dammed lakes at Mount St. Helens (Coldwater and Castle Lakes) by constructing engineered outlet channels. The Spirit Lake drainage tunnel continues to function well, although periodic inspection and maintenance of the tunnel are necessary. None of the stabilized lakes at Mount St. Helens have had outbreaks (Sager and Budai[1989]; Willingham[2005]).

Lahar Diversion

Lahars can be prevented from spreading out and depositing in critical areas by keeping them channelized in modified natural channels or by engineering new channels. Such artificial channels (Figure 7a) must be sufficiently smooth, steep, and narrow (to maintain sufficient flow depth) in order to prevent in-channel deposition. The goal of such channelization is to keep lahars flowing so that they bypass critical areas. The effectiveness of this approach depends on lahar size and composition, channel dimensions, and construction techniques. Highly concentrated lahars (debris flows) can transport large boulders at high velocity and are extremely erosive, so channel bottoms and sides must be lined with concrete or stone masonry surfaces. Even so, hardened diversion channels may require frequent maintenance. Without hardening, lahars in diversion channels can easily erode channel boundaries and establish new flow paths. Channelization of lahar-prone streams draining volcanoes is relatively common in Japan and Indonesia (Smart [1981]; Japan Sabo Assoc. [1988]; Chanson [2004]).

Figure 7: Types of lahar diversion structures. (A) Engineered channel reach in small river draining Sakurajima volcano in southern Japan, where channel is revetted with reinforced concrete and engineered to be as steep, narrow, and smooth as possible, in order to divert lahars away from a developed area. (B) Training dike revetted with steel sheet piles on the lower flank of Usu volcano, Japan and designed to deflect lahars away from buildings and other infrastructure. USGS photos by TCP.

Deflection and diversion structures also can be employed to reroute or redirect lahars away from critical infrastructure or communities. Structures include (a) tunnels or ramps to direct flows under or over roads, railroads, and pipelines; (b) training dikes (also termed levees or bunds) oriented sub-parallel to flow paths to guide lahars past critical areas; and (c) deflection berms oriented at sharper angles to flow paths to force a major course alteration in a lahar (Baldwin et al. [1987]; Hungr et al. [1987]; Huebl and Fiebiger [2005]; Willingham [2005]). However, lahar diversion may cause additional problems (and political resistance) if the diversion requires the sacrifice of only marginally less valuable land. Diversion ramps and tunnels are more practical for relatively small flows, whereas training dikes and deflection berms can be scaled to address a range of lahar magnitudes.

Dikes and berms are constructed typically of locally derived earthen material, but to be effective, these structures must be revetted (armored) on surfaces exposed to highly erosive lahars (Figure 7b). Revetment can be accomplished with thick layers of poured-in-place reinforced concrete, heavy concrete blocks or forms, heavy stone masonry faces or walls, stacked gabions, or steel sheet piles; layers of unreinforced concrete only centimeters thick cannot withstand erosion by large lahars (e.g., Paguican et al. [2009]). However, if a well-revetted dike is overtopped, rapid erosion of the unarmored back side of the dike can quickly cause dike failure and breaching nontheless (Paguican et al. [2009]) (Case study 3). In Japan, where probably more of these structures are constructed than anywhere else in the world, a major design criterion is that their orientation should ideally be less than 45° to the expected attack angle of a lahar to minimize overtopping and erosional damage (Ohsumi Works Office [1995]). Sometimes emergency levees are constructed without revetments, but this usually results in unsatisfactory performance, sometimes with disastrous results (Case study 1).

Case Study 3: Lahar and Sediment Containment and Exclusion Structures

In the months following the May 18, 1980 eruption of Mount St. Helens (Washington, USA), the U.S. Army Corps of Engineers (USACE) built a rock-cored earthen sediment-retention structure (N-1 sediment dam) as a short-term emergency measure to try to hold back lahars and some of the volcanic sediment expected to wash downstream (Willingham[2005]). The structure had two spillways made of rock-filled gabions covered with concrete mortar; it was 1,860 m long and 13 m high, and was located approximately 28 km downstream of the volcano. Neither the upstream nor downstream face of the dam was revetted. Within a month of completion, one of the spillways was damaged by high flow. That spillway was repaired and resurfaced with roller-compacted concrete. In slightly more than a year, the N-1 debris basin filled with about 17 million m3of sediment, and the bed of the river aggraded nearly 10 meters. During the summer of 1981, the USACE excavated 7.4 million m3from the debris basin, but the river replaced that amount and added more during the following winter. The dam was overtopped and breached in quick succession by two events in early 1982—a major winter flood in February and an eruption-triggered, 10-million-m3lahar in March. Overtopping caused deep erosion of the downstream face of the dam at several points, which led to breaching. Even the reinforced, roller-compacted concrete spillways were scoured tens of centimeters, exposing ends of steel reinforcing bars that were abraded to dagger-like sharpness. The extensive damage to the dam and the limited capacity of the catch basin resulted in abandonment of the project (Pierson and Scott[1985]; Willingham[2005]).

Several years later, the USACE started construction of another larger sediment-containment dam (the Sediment Retention Structure or SRS), which was completed in 1989 and further modified in 2012 (Figure8 a). It was built 9 km downstream of the original N-1 structure. In addition to trapping fluvial sediment, it was also designed to intercept and contain a possible future lahar (estimated peak discharge up to 6000 m3/s) from a potential breakout from Castle Lake. The SRS is a concrete-faced (upstream face), rock-cored, earthen dam about 550 m long, 56 m high, 21 m wide at the crest, and has a 122-m-wide armored spillway; its upstream catch basin is 13 km2in area and was designed to hold back about 200 million m3of sediment (USACE— Portland District, unpublished data). By 2005, infilled sediment reached the level of the spillway, and river bed-load sediment began to pass through the spillway, even though the catch basin was filled only to 40% of estimated capacity. After 2005, only a fraction of the river's sediment load was being intercepted, so raising of the spillway by an additional 2.1 m was completed in 2012 and experiments are continuing to induce greater sediment deposition in the upstream basin. The SRS has performed an important function in preventing large amounts of sediment from reaching and filling a reach of the Cowlitz River farther downstream and thus preventing serious seasonal flooding in communities along that river. No attempt has yet been made to excavate and remove sediment from behind the SRS.

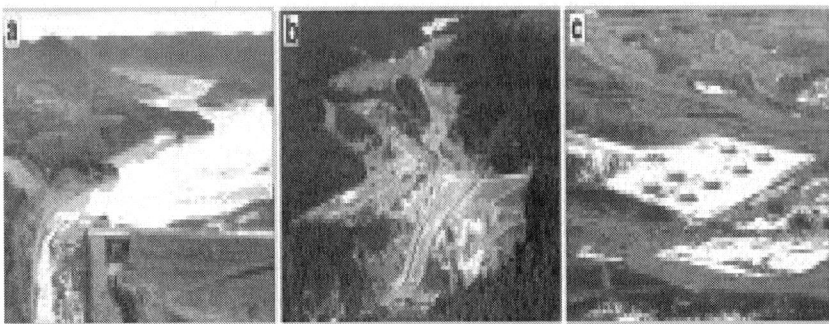

Figure 8: Examples of large-scale lahar containment and exclusion structures. (a) The Sediment Retention Structure (SRS) downstream of Mount St. Helens, USA, built specifically to contain potential lahars and eroded sediment (USGS photo by Adam Mosbrucker, 11 Nov 2012); the volcano is visible on the horizon on the left side of the image. (b) Mud Mountain Dam with a large concrete overflow spillway on the White River downstream of Mount Rainier (USA), (Stein [2001]). It was built as a flood-control structure but it also may function as a trap for at least part of future lahars because little water is normally impounded behind the dam (photo courtesy of U.S. Army Corps of Engineers). (c) Exclusion levees surrounding the Drift River oil terminal on an alluvial plain approximately 40 km downstream of Redoubt Volcano, Alaska (USGS photo by Chris Waythomas, 4 Apr 2009).

An example of a lahar exclusion structure is the levee system enclosing the Drift River Oil Terminal (DROT) in Alaska (USA), which is a cluster of seven oil storage tanks that receive crude oil from Cook Inlet oil wells via a pipeline, plus some buildings and an air strip (Dorava and Meyer[1994]; Waythomas et al.[2013]). The DROT is located on the broad, low-gradient flood plain at the mouth of the Drift River, about 40 km downstream of Redoubt Volcano (Figure8 c). Oil is pumped from these tanks to tankers anchored about 1.5 km offshore at a pumping-station platform. A U-shaped levee enclosure (built around the DROT but open at the downstream end) was raised to a height of 8 m following the 1989–1990 eruption, in order to increase protection of the facility from lahars and flooding. During both the 1989–1990 and 2009 eruptions of Redoubt, lahars were generated that flowed (at low velocity) up against the levees. Minor overtopping of the levees and backflow up from the open end caused some damage and periodic closure of the facility. The river bed aggraded to within 0.5 m of the levee crest in 2009, and the levees were thereafter reinforced and raised higher. The levee enclosure basically did its job, though it would have been more effective if the enclosure had been complete (on four sides).

Lahar Containment or Exclusion

Various structures can prevent lahars from reaching farther downstream, or seal off and protect critical areas while surrounding terrain is inundated. Sediment retention dams (Figure 8a) or containment dikes are used hold back as much sediment as possible but not necessarily water. To contain lahars, they must be constructed to withstand erosion and possible undercutting along their lateral margins and be tall enough to avoid overtopping. Under-design of these structures or inadequate removal of trapped sediment behind them can result in eventual overtopping and failure of the structure (e.g., Paguican et al. [2009]; Case study 3). The area upstream of a barrier where sediment is intended to accumulate is usually termed the catch basin or debris basin. Small excavated catch basins are also termed sand pockets. Such accumulation zones are typically designed to accommodate sediment from multiple flow events, and large tracts of land may be needed for this purpose. However, acquisition of land for this purpose can be problematic (Case study 1). If the design capacity is not large enough to accommodate all of the sediment expected to wash into a catch basin, provisions must be made to regularly excavate and remove accumulated sediment.

In addition to specially built lahar-related structures, pre-existing dams can sometimes be useful in containing all or most of the debris in a lahar (Figure 8b). Dams built for flood control or for impoundment of water for hydroelectric power generation or water supply can contain lahars and prevent them from reaching downstream areas, as long as (a) sufficient excess storage capacity exists behind the dam to accommodate the lahar volume, and (b) there is no danger of lahar-induced spillover at the dam in a way that could compromise dam integrity and lead to dam failure. Reservoir drawdown during volcanic activity might be necessary to ensure sufficient storage capacity to trap a lahar. This was done at Swift Reservoir on the south side of Mount St. Helens prior to the 1980 eruption, allowing it to successfully contain two lahars totaling about 14 million m^3 (Pierson [1985]).

Exclusion dikes can enclose and protect valuable infrastructure, as was done in 1989–1990 and 2009 to protect oil storage tanks at the mouth of the Drift River, Alaska, from lahars and volcanic floods originating from Redoubt Volcano (Dorava and Meyer [1994]; Waythomas et al. [2013]) (Case study 3; Figure 8c). Diked enclosures may be a more appropriate strategy than channelization, diversion, or deflection in areas with low relief where low channel gradients encourage lahar deposition and where areas to be protected are small relative to the amount of channelization or diking that otherwise would be required.

Check Dams to Control Lahar Discharge and Erosion

Some structures are built to slow down or weaken lahars as they flow down a channel. Check dams are low, ruggedly built dams that act as flow impediments in relatively steep stream channels (Figures 9 and 10). They have four functional roles: (a) to prevent or inhibit downcutting of the channel, which in turn inhibits erosion and entrainment of additional sediment; (b) to trap and retain some of a lahar's sediment, thereby decreasing its volume; (c) to add drop structures to the channel profile in order to dissipate energy and slow downstream progress of the lahar; and (d) to induce deposition in lower-gradient reaches between dams (Smart [1981]; Baldwin et al.[1987]; Hungr et al. [1987]; Johnson and McCuen [1989]; Armanini and Larcher [2001]; Chanson[2004]; Huebl and Fiebiger [2005]; deWolfe et al. [2008]).

Figure 9: Examples of permeable lahar flow-control structures. (a)Steel-pipe slit dam at Mount Unzen, Japan. (b) Drain-board screen at Mount Yakedake, Japan, after having stopped the bouldery head of a small debris-flow lahar. USGS photos by TCP.

Figure 10: Examples of impermeable lahar flow- and erosion-control structures. (a) Series of sheet-pile check dams with masonry aprons at Mount Usu, Japan. (b) Dam of rock-filled steel cribs at Mount Ontake, Japan. USGS photos by TCP.

Check dams are commonly built in arrays of tens to hundreds of closely spaced dams that give a channel a stair-step longitudinal profile. Very low check dams are also called stepped weirs and are commonly constructed between larger check dams to act as hydraulic roughness elements for large flows (Chanson [2004]). A variety of styles and sizes of check dams have been developed, but fall into two basic categories: permeable or impermeable.

Permeable slit dams, debris racks, and open-grid dams (Figure 9a) are constructed of heavy tubular steel or structural steel beams, commonly with masonry bases and wing walls. Such structures are designed to act as coarse sieves, catching and retaining boulder-size sediment in a lahar but allowing finer material and water to pass through with depleted energy and mass. In addition to reducing the velocity of flows as they pass through, these dams also attenuate peak discharge. The effect is most pronounced on granular (clay-poor) debris-flow lahars that typically have steep, boulder-laden flow fronts. A variation on these vertically oriented structures is the drain-board screen (Azakami [1989]) (Figure 9b), which is a horizontally oriented steel grate or grill that performs the same sieving function for boulders as permeable dams when a lahar passes over the top of the grate, retaining coarse clasts while water and finer sediment drop down through the grate. Because of their orientation, these structures do not have to withstand the same high lateral forces as the upright permeable dams.

Impermeable check dams are composed of solid concrete, concrete with a packed earthen core, or steel cribs or gabion baskets filled with rocks and gravel (Figure 10). They may have small slits or pipes to allow exfiltration of water through the dam, in order to minimize impoundment of water. Gabions are used widely in the developing world because of their low construction costs—gravel fill often can be excavated locally from the channel bed, their permeability, and their flexibility, which can allow a dam to sag without complete failure if undermined by erosion. The crests of impermeable check dams commonly slope toward the center of the dam, where a notch or spillway is constructed, in order to direct streamflow or lahars over the dam onto a thick concrete apron extending downstream to protect the toe of the dam from erosion. Concrete sills or roughness elements commonly are placed at the downstream ends of aprons to further slow the flow that passes over the main dam. If upstream catch basins fill to capacity with sediment, check-dam functions are then limited to a, c, and d noted above, but full functionality can be restored if catch basins are regularly excavated.

Hazard Warning

Where communities already occupy lahar hazard zones or where transient populations move in and out, a lahar warning system can be an option that would allow an at-risk population to safely evacuate prior to lahar arrival, whether or not used in conjunction with engineered protection structures. Lahar warning systems can minimize fatalities, but they are not practical in every situation. In cases where populations are situated close to a lahar source area, there simply may be little or no time for a timely warning to be issued and for people to receive it in time to evacuate (Cardona [1997]; Pierson [1998]; Leonard et al. [2008]). Timing is even more challenging at volcanoes where lahars unrelated to ongoing or recent volcanic activity can occur— where volcanic edifices are weakened by hydrothermal alteration, for example, because lahar occurrence generally would not be anticipated. The decision of whether or not to install a warning system should also consider the long-term and ongoing needs for sustaining coordination and communication among the many organizations and individuals involved, regularly maintaining and testing the instrumentation, and keeping at-risk populations informed and prepared, especially where populations are transient.

Lahar warning systems have three basic components: (1) sensors or observers to detect an approaching lahar; (2) data acquisition, transmission, and evaluation systems to transfer and evaluate data to determine if there really is an approaching lahar; and (3) alert-notification systems to inform people that a lahar is coming. The spectrum of ways to accomplish these functions can range from simple 'low-tech' approaches largely involving human observers to more sophisticated 'high-tech' systems (Figure 11). In addition to these basic components that warn of an approaching lahar, integrated (often called "end-to-end") warning systems also include components that not only warn people but prepare them and lead them to respond proactively and to assume personal responsibility for evacuating. These additional components include pre-event planning and preparation; mechanisms to formulate and target appropriate warning messages; effective outreach to at-risk populations so that they understand what to do when a warning is received; establishment of evacuation routes and safe refuges that can be reached (generally on foot) before lahar arrival; and evacuation exercises with follow-up evaluation (Mileti and Sorenson[1990]; Basher [2006]; Leonard et al. [2008]).

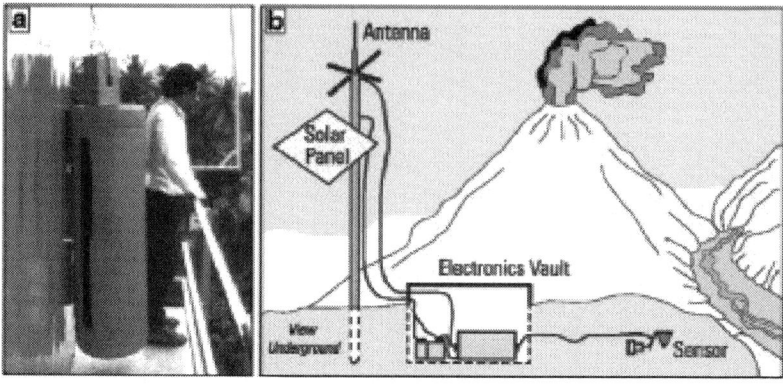

Figure 11: Examples of "low-tech" and "high-tech" lahar detection systems. (a) Human observer in lahar observation tower along a river that originates on Merapi volcano, Indonesia; observer strikes the large hanging steel drum ("tong-tong") with a steel bar after seeing or hearing an approaching lahar. USGS photo by TCP. (b) Schematic diagram of an acoustic flow monitor (AFM)—a sensor that detects ground vibrations generated by an approaching lahar, then telemeters that information in real time to a base station, where the signal is evaluated and a decision is made on whether or not to issue an alarm (see LaHusen [2005]).

Once a warning system becomes operational and depended upon, there must be sufficient ongoing funding and institutional commitment to continue operation indefinitely and to regularly educate and train the at-risk population. This is important because termination of a warning system while the hazard still exists may involve liability and ethical issues. Long-term operation costs include not only those for the normal maintenance of warning-system components, but also replacement costs if components are vandalized or stolen and, where necessary, costs for providing instrument-site security.

Volcano scientists play important roles, not only in developing or deploying warning system instrumentation, but also in training emergency managers to confidently interpret scientific and technical information from the monitoring systems. Scientists also can help to develop clear warning messages that are appropriate and understandable by affected populations (Mileti and Sorenson [1990]). Although lahar warning systems can issue false alarms, research shows that the "cry wolf" syndrome does not develop within affected populations as long as people understand the hazard and are later told about the possible reasons why a false warning was issued (Mileti and Sorenson [1990]; Haynes et al. [2008]).

'Low-Tech' Warning Systems

In some developing countries, effective low-tech warning systems employ human observers to alert threatened populations. Observers can be positioned at safe vantage points within view of lahar-prone river channels at times when flows have a high likelihood of occurring, such as during ongoing eruptions and during and following intense rainfall, particularly within the first few years after eruptions (de Bélizal et al. [2013]; Stone et al. [2014]). Observers stationed near lahar source areas are in a position to see or hear localized convection-cell rain storms that can trigger lahars, and human hearing can be very effective in detecting the approaching lahars themselves, often minutes before they come into view. The low-frequency rumbling sound caused by large boulders grinding against the river bed can carry hundreds or thousands of meters through the air and through the ground—a sound that is unmistakable to a trained observer. For example, a relatively small lahar occurring recently at Mount Shasta, California, sounded "like a freight train barreling down the canyon" and at times "like a thunder rumble" to a U.S. Forest Service climbing ranger (Barboza [2014]).

Once a lahar is detected, an observer can quickly issue an alert directly (by drum, siren, cellular phone, hand-held radio, etc.) to people living nearby (Figure 11a). This basic approach to lahar detection may be preferable where there is limited technical or financial capacity for maintaining sensors and other electronic equipment, where there are safe and accessible observation points, where there is high likelihood of expensive instruments being damaged or stolen without someone to guard them, where environmental conditions are challenging, or where electrical power and telecommunications are unreliable. Lahar detection by human observers is not immune to failure, however. Reliability is a function of the trustworthiness and alertness of the observers, their level of training, and the effectiveness of the alert notification method.

Automated Telemetered Warning Systems

Automated electronic warning systems can be used to detect approaching lahars and telemeter alerts in areas where electrical power, technical support capabilities, and funding are more assured. Systems also can be designed to detect anomalous rainfall or rapid snowmelt that could trigger lahars, sense incipient motion of an unstable rock mass or lake-impounding natural dam, or detect an eruption that could trigger a lahar (Marcial et al. [1996]; Sherburn and Bryan [1999]; LaHusen [2005]; Manville and Cronin [2007]; Leonard et al. [2008]; USGS [2013]) (Figure 11b). In order for data from any of these various sensors to be useful for alert notification, they must be transmitted from remote sites in real time to a receiving station. Transmission can be accomplished by either ground-based or satellite-based radio telemetry (LaHusen [2005]) or cellular phone (Liu and Chen [2003]). Alert notifications can occur either automatically when some threshold in the level of the detection signal is exceeded, or an intermediate step can involve emergency management personnel, who verify and validate the detection signal before an alert is issued. Coordination among multiple agencies is critical to the success of an automated system, because hardware and software development of the sensor and the data acquisition/transmission systems are typically handled by physical scientists and engineers, whereas the development, operation, and maintenance of warning systems are typically managed by emergency managers and law-enforcement personnel (Case study 4).

Case Study 4: the Mount Rainier Lahar Warning System

A significant volume of rock on the upper west flank of Mount Rainier (USA) has been extensively weakened (60–80% loss in unconfined strength) by hydrothermal alteration and is unstable (Watters et al. [2000]; Finn et al.[2001]; John et al.[2008]). A lahar warning system was developed by the U.S. Geological Survey and Pierce County (Washington) to detect potential lahar initiation from this sector, and it was installed in 1995 by USGS and Pierce County personnel in the Carbon and Puyallup River valleys downstream of the weak and oversteepened rock mass (USGS[2013]). The system is designed to warn tens of thousands of people who live in the downstream lahar hazard zone of an approaching lahar. Affected communities are situated from 40 to 80 km downstream of the volcano and could have from 12 minutes to 2 hours, depending on location, to evacuate after receiving a warning message. Since installation, the warning system has been maintained and operated by the Pierce County Department of Emergency Management, in collaboration with the Washington State Emergency Management Division.

The system comprises specialized seismic sensors capable of detecting ground vibrations within a frequency range typical of lahars (30–80 Hz), a ground-based radio telemetry system for detection-signal transmission, and a combination of sirens, direct notification, and the Emergency Alert System (EAS) that utilizes NOAA weather radios for warning message dissemination (LaHusen[2005]; USGS[2013]). County and state emergency-management agencies and city and county law-enforcement agencies collectively have responsibility for verifying and validating alerts from the sensors, activating warning sirens, and sending warning messages.

Collaboration between all the agencies involved in lahar hazard warning and risk reduction at Mount Rainier is fostered by regular meetings of the "Mount Rainier Work Group". Such lahar warning systems require ongoing collaboration between scientists and emergency management officials, as well as regular maintenance and testing. Members of the at-risk population (including schools) have been assigned evacuation routes, have been informed about what to do when a warning message is received, and regularly participate in evacuation drills (Figure 3g).

Warning Message Development and Delivery

In the simplest warning systems, warning messages are delivered only as simple audible signals (drums, sirens, whistles, etc.), and the affected population must be informed beforehand about what the signals mean and what the appropriate response should be. In more sophisticated systems, incident-specific alert messages can be delivered to large populations simultaneously by cellular phone, the Internet, radio, or television. In these cases, the alert must convey a definitive and unambiguous message that effectively prompts individuals to take protective actions. Several factors influence the effectiveness of a warning message, including the content and style of the message, the type and number of dissemination channels, the number and pattern of warning statements, and the credibility of the warning source (Mileti and Sorenson [1990]).

Warning messages should be specific, consistent, certain, clear, and accurate (Mileti and Sorenson[1990]). To ensure credibility, message content should include a description of the hazard and how it poses a threat to people, guidance on what to do to maximize personal safety in the face of impending danger, location of the hazard, the amount of time people have to take action, and the source of the warning. The more specific a warning message is, the more likely the receiver is to accept the warning (Cola [1996]; Greene et al. [1981]). Emergency warnings without sufficient detail create information voids, and the affected population may then rely on ill-informed media commentators, friends, neighbors, or personal bias and perceptions to fill this void (Mileti and Sorenson [1990]). Input from volcano scientists is critical for some of this detail and specificity.

Both credibility and consistency of the warning message are important. At-risk populations commonly receive information from informal sources (for example, the media, friends, social media), sometimes more quickly than through various official channels during a crisis (Mileti[1999]; Leonard et al. [2008]; Dillman et al. [1982]; Mileti and Sorenson [1990]; Parker and Handmer [1998]; Mei et al. [2013]). For example, 40–60% of people in the vicinity of Mount St. Helens first received informal notification of the 1980 eruption (Perry and Greene [1983]; Perry[1985]). The proliferation of informal information channels today with the Internet and social media can benefit the warning dissemination process, because individuals are more likely to respond to a warning if it is confirmed by multiple sources (Cola [1996]; Mileti and Sorenson[1990]). But multiple sources become problematic if they advance conflicting information, causing individuals to become confused. Therefore, challenges for emergency managers and scientists are to keep reliable information flowing quickly and to maintain consistent messages, both during and after an emergency. Joint information centers can ensure that (a) there is consistency in official warning statements among multiple scientific and emergency-management agencies, (b) easy access is provided for the media to the official information and to experts who can explain it, and (c) the effectiveness of warning messages is monitored (Mileti and Sorenson [1990]; Driedger et al. [2008]).

Evacuation Training

Warnings are given so that people in a lahar flow path can move quickly out of harm's way. Sheltering in place is generally not a viable option. The lives of at-risk individuals may depend on understanding that they are living in, working in, driving through, or visiting a lahar hazard zone, as well as understanding what to do when they receive a warning (Mileti and Sorenson [1990]; Leonard, et al. [2008]). As the world witnessed in the 1985 Nevado del Ruiz disaster (Voight[1990]) (Case study 5), warnings that a lahar was bearing down on their town were not able to prevent catastrophic loss of life, because the warnings were issued without the population's understanding of the risk or how they should respond. To increase the likelihood of successful evacuations, scientists should encourage and help lead hazard-response exercises and evacuation drills, especially in areas with short time windows for evacuating hazard zones. These exercises and drills provide emergency managers the opportunity to identify weaknesses in the warning–evacuation process and to minimize potential delays that could result from confusion, insufficient information, or lack of understanding on what to do. They also provide scientists with a platform for discussing past catastrophes and the potential for future events. Holding an annual table-top exercise or community-wide evacuation drill on the anniversary of a past disaster can help to institutionalize and personalize the memory of past events, an important step if new community members are to take these threats seriously. A well-educated and trained community that possesses information about where they will get information and what emergency actions to take is less likely to be confused by warning messages, to resist evacuation orders, or to blame officials for ordering an evacuation when a catastrophic event fails to occur (e.g., Cardona [1997]). The goal for scientists and emergency managers is to create a "culture of safety" (cf., Wisner et al. [2004], p. 372) where at-risk individuals understand potential hazards, take personal responsibility for reducing their risks, understand how to respond to an event, and realize that lessening of risks requires actions from all levels of a community and government.

Case Study 5: the Nevado Del Ruiz Disaster

The 1985 Nevado del Ruiz lahar disaster, which cost approximately 21,000 lives in the town of Armero, Colombia (Figure1 a), is an excellent case study of the complexities that can lead to ineffective evacuation after warning messages are broadcast, poor emergency response, and a haphazard disaster recovery (Voight[1990]; Hall[1992]). In post-event analyses, it was generally concluded that the Ruiz catastrophe was the result of cumulative human and bureaucratic errors, including lack of knowledge, misunderstanding and misjudgment of the hazard, indecision, and even political barriers to effective communication, rather than inadequate science or technical difficulties. Other factors contributing to the catastrophe included evacuation plans that had been prepared but not shared with the public, poorly equipped emergency management authorities, the absence of agreed-upon decision-making processes, and uncertainty about the pre-event hazard assessments that made public officials reluctant to issue an early evacuation order because of the potential economic and political costs. The hazard maps produced by scientists for Nevado del Ruiz prior to the eruption were highly accurate in their predictions of where lahars could go, but they were published only about a month before the disaster, giving little time for assimilation and responsive action by the emergency managers. Furthermore, production of the maps did not lead to effective risk communication, because the scientists who made the maps generally did not engage in conveying that risk information in understandable terms to officials and the public. Scientists may prepare excellent hazard assessments and maps, but unless they participate fully in conveying hazard information to officials and the public in ways that are understandable, disasters can still happen (Voight[1990]; Hall[1992]).

Hazard Response and Recovery Planning

The first three risk-reduction strategies focus on minimizing losses through actions taken before a lahar occurs, but this fourth strategy determines the effectiveness of the immediate emergency response and the longer-term course of recovery after a lahar has occurred, which together define a community's resilience. Hazard response includes the rescue, emergency care, sheltering, and feeding of displaced persons, which is facilitated by a robust incident command system. Such a system could range from coordinated communication in a small village to a structured multi-agency protocol, such as NIMS (National Incident Management System) in the United States (FEMA[2014]). Recovery involves the reestablishment of permanent housing, infrastructure, essential services, and economic viability in the community.

Response to a lahar that has impacted a populated area can be difficult. Lahars present first responders, search-and-rescue teams, and disaster-management officials with challenges unlike some other disasters: (a) the area of impact can be extensive and locally covered by debris from crushed buildings and other structures; (b) the degree of impact is generally greatest toward the center of the impact zone and less along the edges; (c) lahars can transport victims and structures long distances from their initial locations; (d) survivors may be difficult to locate; (e) fresh lahar deposits commonly stay liquefied (like quicksand) for days to weeks, and upstream river flow may cut through a debris field, so that access to victims may be limited to hovering helicopters, small boats, or rescuers on the ground being confined to walking on logs or sheet of plywood (Figure 12); (f) once located, victims can be difficult to extract from the mud; and (g) critical facilities (hospitals, police and fire stations, etc.) may be inaccessible, damaged, or destroyed. These challenges can be critical, because the time window is small for getting injured victims to medical care, and uninjured victims trapped in liquefied mud can quickly become hypothermic. To minimize fatalities from a lahar, communities in hazard-prone areas should develop realistic rescue and response plans that are understood by all individuals and responsible agencies. In addition to developing search and rescue tactics, such plans should include identification of refuge zones, logistical resources, emergency social services, and security personnel that will be needed to establish emergency shelters and for survivors at those shelters, and for site access control and security (see UNDRO [1985], for an emergency plan example). Scientists can support emergency managers and public officials in the aftermath of a catastrophic event by assessing the likelihood of future lahars and floods, the suitability of areas for relief operations, and the evolving stability of lahar deposits.

Figure 12: Examples of challenges to rescue and recovery where thick lique-fied mud and debris have flowed into a populated area—the Highway 530 (Oso, Washington) landslide disaster of 22 March 2014. Soft mud can pre-clude rescue of victims by responders on the ground, particularly in the first hours or days following a lahar. (a) Rescuer being lowered by helicopter to an area where ground is too soft to reach on foot (copyrighted AP photo by Dan Bates, used with permission). (b) Rescuer searching for victims using an inflatable boat, because flooding from backed-up river inundated part of the debris field (copyrighted AP photo by Elaine Thompson, used with permis-sion).

Proper shelter planning is critical to minimize the potential for additional victims. Poor planning of emergency shelters and camps can create new disaster victims due to disease outbreaks and malnutrition if shelter is inadequate and timely supply of food, clean water, and medicine does not occur. Shelter planning should also take into account the quality of life and livelihoods for displaced populations. For example, 50 to 70% of people displaced by the 2010 eruption of Mt. Merapi (Indonesia) ignored evacuation orders and consistently returned (in some cases daily) to danger zones during the crisis because of the need to care for livestock and to check on possessions (Mei et al. [2013]). The lack of activities and work programs in the evacuation camps also can result in people leaving the shelters. In addition, if schools are used as shelters, then public education suffers because school buildings are occupied by evacuees. In countries with limited relief resources, people may be better served if extended families can temporarily house impacted relatives during emergencies. Community leaders, with assistance from scientists, can encourage residents to develop their own evacuation and relocation strategies.

Following an initial disaster response, recovery becomes the next goal. Restoring community functions is typically a top priority in the aftermath of an extreme event such as a lahar, but quick reconstruction may not be possible if key infrastructure, industrial parks, downtown cores of communities, and extensive areas of residential housing are buried or swept away (Tobin and Whiteford [2002]). Pre-event recovery planning, however, can allow resilient communities to recover more quickly by prioritizing the building of redundant and diversified back-up systems, services, and infrastructure into their communities beforehand. For transportation networks for example, this could mean having multiple routes to critical or essential facilities, predetermined appropriate sites for helipads or temporary airstrips, and storage sites for heavy equipment—all located outside of the hazard zone. Scientists can assist the development of recovery plans by providing advice on where future commercial, residential, and industrial districts could be located outside of hazard zones. A well thought-out recovery plan also provides an impacted community with opportunities for the established social fabric of a community to be maintained, for relocation to a safer site, and for comprehensive redevelopment that avoids haphazard or fragmented future growth.

Resettlement following a disaster is not simply a matter of rebuilding homes and infrastructure at a safer site. The quality of life, means of making a living, and social needs and networks of displaced populations must be recognized for resettlement to be successful, and residents must be part of the planning process. For example, Usamah and Haynes ([2012]) document low occupation rates of (and minimal owner investment in) government-provided housing at permanent relocation sites two years after the Mayon volcano (Philippines) eruption in 2006. They attribute this to the lack of community planning participation, lack of appreciation of original house design and function (for example, metal roofs on new houses make them hotter during the day than traditional houses with palm-thatch roofs), delays in utility infrastructure, no public facilities such as religious centers and schools, few livelihood options, and little long-term community development. Although authorities and donors (and residents) were satisfied that the new housing was safer, interviewees felt the long-term objective of facilitating sustainable lives was ignored. A similar reluctance to participate in a resettlement program was found at Colima volcano (Mexico) for many of the same reasons (Gavilanes-Ruiz et al. [2009]).

Thus, community participation in long-term recovery planning is needed to ensure identification of the community's needs and the community's support.

Development of an effective recovery plan can ensure provision of a number of practical recovery needs. Those needs include: achievement of more appropriate land-use regulations, identification of funding sources for reconstruction, identification of resources and disposal sites for debris clearance, enlistment of economic support for recovering businesses, and adoption of new construction standards. Recovery plans help ensure that reconstruction after the event does not reoccupy a hazard zone or happen in an ad hoc fashion. Scientists can contribute to this planning process by (a) helping public officials visualize the probable physiographic, geologic, and hydrologic realities of a post-event landscape; and (b) identifying what post-event hazards would be relevant for the community.

SCIENTIST ROLES IN LAHAR RISK REDUCTION

All four of the basic strategies for lahar-hazard risk reduction—hazard avoidance, modification, warning, and response/recovery—require the input and judgment of volcano scientists, even though emergency managers and public officials have the responsibility for their planning and implementation. In addition, scientists play a critical role in educating emergency managers, public officials, and at-risk populations about lahar hazards. Specific ways that scientists can participate are discussed in the sections above.

Some scientists are uncomfortable participating in processes that are influenced (if not dominated) by social, economic, and political factors. However, risk managers cannot successfully manage natural threats to communities without involvement by scientists (Peterson [1988], [1996]; Hall[1992]; Haynes et al. [2008]). Peterson ([1988]) goes as far to say that scientists have an ethical obligation to effectively share their knowledge to benefit society by making their knowledge understandable to non-scientists. Scientists can communicate hazard information to the public through formal and informal face-to-face meetings, through public presentations, and through the media.

Qualities exhibited by scientists that enhance their trustworthiness in the eyes of the public are reliability (consistency and dependability in what they say), competence (having the skills and ability to do the job), openness (having a relaxed, straightforward attitude and being able to mix well and become 'part of the community'), and integrity (having an impartial and independent stance) (Pielke [2007]; Haynes et al. [2008]). Yet there is always a potential for friction and other distractions during the stressful time of a volcano crisis, and scientists should recognize and try to avoid the various problems related to personal and institutional interactions that have plagued the credibility of scientists during past volcanic crisis responses, such as communications breakdowns and disputes among scientists (with different messages coming from different scientists), scientists advocating for particular mitigation strategies, scientists avoiding or "talking down" to the public, poor scientific leadership, failure to recognize cultural differences between themselves and affected populations, and failure to share information and scarce resources (Newhall et al. [1999]).

Effective lahar-hazard risk reduction cannot occur unless the hazard and its attendant risks are recognized by authorities and the public, and this recognition is affected by the willingness and ability of scientists to communicate hazards information (Peterson [1988]). The contributions of scientists will be effective if they are willing to embrace their educational, interpretive, and advisory roles, to work in partnership with officials and the public, and to be sensitive to the cultural norms of the society in which they are working. Scientists must be willing and able to participate in community events, hone skills related to public speaking, work with the media, and work one-on-one with community leaders. As Newhall et al. ([1999]) state, the guiding principle for scientists during volcanic crises should be to promote public safety and welfare. This principle extends to non-crisis situations, as well, and scientists can and should work with officials and the public frequently to lessen the risk from future lahars. In short, lahar-hazard risk reduction cannot be effectively accomplished without the active, impartial involvement of qualified scientists.

AUTHORS' CONTRIBUTIONS

TCP developed the risk-reduction strategy categorization, evaluated the effectiveness of many of the strategies, and wrote approximately

60% of the manuscript. NJW summarized the ways that community vulnerability and risk reduction can be applied to lahar hazards, and he wrote approximately 40% of the manuscript. CLD edited and revised an early draft of the section on hazard and risk education. This section is in large part a distillation of CLD's findings on how hazards information can be communicated effectively and understandably. All authors read and approved the final manuscript.

ACKNOWLEDGMENTS

This review of strategies for lahar risk reduction is based not only on the literature cited, but also on observations made by the authors of the practical application of these techniques in many parts of the world, combined with their own direct experience and research. Photographs with credits in the form of initials were taken by the authors. Work by the authors on this topic has been supported over the years by the USGS Volcano Hazards Program, the USGS/USAID–OFDA Volcano Disaster Assistance Program, and the USGS Land Change Science Program. We thank Kelvin Rodolfo, Franck Lavigne, and one anonymous reviewer for their insightful reviews of an earlier version of the article. Any use of trade, product, or firm names is for descriptive purposes only and does not imply endorsement by the U.S. Government.

REFERENCES

1.   Armanini A, Larcher M (2001) Rational criterion for designing opening of slit check dam. J Hydraul Eng ASCE 127:94-104

2.   Aucker M, Sparks R, Siebert L, Crosweller H, Ewert J (2013) A statistical analysis of the global historical volcano fatalities record. J Appl Volcanol 2(2):24

3.   Azakami S (1989) Recent trends in sabo control measures in Japan. In: Proceedings of the International Symposium on Erosion and Volcanic Debris Flow Technology. Ministry of Public Works, Yogyakarta, Indonesia. pp KO2-1-KO2-9

4.   Baldwin JE, Donley HF, Howard TR (1987) On debris flow/avalanche mitigation and control, San Francisco Bay area, California. In: Costa JE, Wieczorek GF (eds) Debris Flows/

Avalanches: Process, Recognition, and Mitigation, Geol Soc Am Reviews in Engineering Geology, pp. 223-236

5. Barboza T (2014) Forest Service Thinks California's Drought Caused a Massive Mudslide. Los Angeles Times (Science section).

6. Basher R (2006) Global early warning systems for natural hazards: systematic and people-centred. Phil Trans Royal Soc A 364:2167-2182

7. Beyers JL (2004) Postfire seeding for erosion control: effectiveness and impacts on native plant communities. Conserv Biol 18:947-956

8. Bird D, Gisladottir G, Dominey-Howes D (2010) Volcanic risk and tourism in southern Iceland: Implications for hazard, risk and emergency response education and training. J Volcanol Geotherm Res 18:33-48

9. Cardona O (1997) Management of the volcanic crises of Galeras volcano: social, economic and institutional aspects. J Volcanol Geotherm Res 77:313-324

10. Chanson H (2004) Sabo check dams—mountain protection systems in Japan. Internat J River Basin Mgmt 2:301-307

11. Chenet M, Grancher D, Redon M (2014) Main issues of an evacuation in case of volcanic crisis: social stakes in Guadeloupe (Lesser Antilles Arc). Nat Hazards 73:2127-2147

12. Chester D (2005) Theology and disaster studies: the need for dialogue. J Volcanol Geotherm Res 146:319-328

13. Cola R (1996) Responses of Pampanga households to lahar warnings--lessons from two villages in the Pasig-Potrero River watershed. In: Newhall CG, Punongbayan RS (eds) Fire and Mud: Eruptions and Lahars of Mount Pinatubo, Philippines, University of Washington Press, Seattle, WA. pp 141-149

14. Collins BD, Dunne T (1986) Erosion of tephra from the 1980 eruption of Mount St. Helens. Geol Soc Am Bull 97:896-905

15. Crittenden K (2001) Can this town survive? Case study of a buried Philippine town. Nat Hazards Rev 2:72-79

16. Crittenden K, Rodolfo K (2002) Bacolor Town and Pinatubo Volcano, Philippines: coping with recurrent lahar disaster. In: Torrence R, Grattan J (eds) The Archaeology of Natural Disasters, One World Archaeology Series, Routledge, London. pp 43-65

17. Cronin S, Petterson M, Taylor P, Biliki R (2004) Maximising multi-stakeholder participation in government and community volcanic hazard management programs: a case study from Savo, Solomon Islands. Nat Hazards 33:105-136

18. Cronin SJ, Gaylord DR, Charley D, Alloway BV, Wallez S, Esau JW (2004) Participatory methods of incorporating scientific with traditional knowledge for volcanic hazard management on Ambae Island, Vanuatu. Bull Volcanol 66:652-668

19. Cummans J (1981) Chronology of mudflows in the South Fork and North Fork Toutle River following the May 18 eruption. In: Lipman PW, Mullineaux DR (eds) The 1980 eruptions of Mount St, US Geol Surv Professional Paper 1250, Helens, Washington. pp 479-486

20. de Bélizal E, Lavigne F, Robin AK, Sri Hadmoko D, Cholik N, Thouret JC, Sawudi DS, Muzani M, Sartohadi J, Vidal C (2013) Rain-triggered lahars following the 2010 eruption of Merapi volcano, Indonesia: A major risk. J Volcanol Geotherm Res 261:330-347

21. deWolfe VG, Santi PM, Ey J, Gartner JE (2008) Effective mitigation of debris flows at Lemon Dam, La Plata County, Colorado. Geomorphol 96:366-377

22. Dillman DA, Schwalbe ML, Short JF (1982) Communication behavior and social impacts following the May 18, 1980 eruption of Mount St. Helens. In: Keller SAC (ed)   , Three Years Later, Pullman, WA, Washington State University, Mt. St. Helens. pp 173-179

23. Dominey-Howes D, Minos-Minopoulos D (2004) Perceptions of hazard and risk on Santorini. J Volcanol Geotherm Res 137:285-310

24. Dorava JM, Meyer DF (1994) Hydrologic hazards in the lower Drift River basin associated with the 1989–1990 eruption of Redoubt Volcano, Alaska. In: Miller TP, Chouet BA (eds) The 1989–1990 Eruptions of Redoubt Volcano, Alaska, pp. 387-407

25. Driedger CL, Scott WE (2008) Mount Rainier—Living Safely With a Volcano in your Backyard.

26. Driedger CL, Scott WE (2010) Volcano hazards. In: Schelling J, Nelson D (eds) Media guidebook for natural hazards in

Washington—addressing the threats of tsunamis and volcanoes, Washington Military Department Emergency Management Division,

27. Driedger CL, Wolfe EW, Scott KM (1998) Living With a Volcano in your Back Yard: Mount Rainier Volcanic Hazards—A Prepared Presentation for use by Public Officials and Educators.

28. Driedger C, Stout T, Hawk J (2002) The Mountain is a Volcano!— Addressing geohazards at Mount Rainier. J Assoc Nat Park Rangers 18:14-15

29. Driedger C, Doherty A, Dixon C, Faust L (2014) Living with a volcano in your backyard—An educator's guide with emphasis on Mount Rainier (ver. 2.0, December 2014).

30. Driedger CL, Neal CA, Knappenberger TH, Needham DH, Harper RB, Steele WP (2008) Hazard information management during the autumn 2004 reawakening of Mount St. Helens volcano, Washington. In: Sherrod DR, Scott WE, Stauffer PH (eds) A Volcano Rekindled: The Renewed Eruption of Mount St. Helens, 2004–2006. US Geol Surv Professional Paper 1750, pp. 505-519

31. Driedger CL, Westby L, Faust L, Frenzen P, Bennett J, Clynne M (2010) 30 cool facts about Mount St. Helens.

32. Dzurisin D, Driedger CL, Faust L (2013) Mount St. Helens, 1980 to Now—What's Going On? US Geol Surv Fact Sheet 2013–3014.

33. (2014) National Incident Management System. U.S. Federal Emergency Management Agency.

34. Finn CA, Sisson TW, Deszcz-Pan M (2001) Aerogeophysical measurement of collapse-prone hydrothermally altered zones at Mount Rainier volcano. Nature 409:600-603

35. Frenzen PM, Matarrese MT (2008) Managing public and media response to a reawakening volcano: lessons from the 2004 eruptive activity of Mount St. Helens. In: Sherrod DR, Scott WE, Stauffer PH (eds) A Volcano Rekindled: The Renewed Eruption of Mount St. Helens, 2004–2006. US Geol Surv Professional Paper 1750, pp. 493-503

36. Gaillard J, Maceda A (2009) Participatory three-dimensional mapping for disaster risk reduction. Participatory Learn Action 60:109-118

37. Gardner CA, Guffanti MC (2006) US Geological Survey's Alert Notification System for Volcanic Activity.

38. Gardner CA, Scott WE, Major JJ, Pierson TC (2000) Mount Hood—History and Hazards of Oregon's Most Recently Active Volcano.

39. Gavilanes-Ruiz JC, Cuevas-Muñiz A, Varley N, Gwynne G, Stevenson J, Saucedo-Girón R, Pérez-Pérez A, Aboukhalil M, Cortés-Cortés A (2009) Exploring the factors that influence the perception of risk: The case of Volcán de Colima, Mexico. J Volcanol Geotherm Res 186:238-252

40. Gray DH, Sotir RB (1996) Biotechnical and Soil Bioengineering Slope Stabilization: a Practical Guide for Erosion Control. Wiley-Interscience, New York.

41. Greene M, Perry R, Lindell M (1981) The March 1980 eruptions of Mt. St. Helens: citizens' perceptions of volcano threat. Disasters 5:49-66

42. Gregg C, Houghton B, Johnston D, Paton D, Swanson D (2004) The perception of volcanic risk in Kona communities from Mauna Loa and Hualalai volcanoes, Hawai'i. J Volcanol Geotherm Res 130:179-196

43. Guadagno FM, Revellino P (2005) Debris avalanches and debris flows of the Campania Region, southern Italy (Chapter 19). In: Jakob M, Hungr O (eds) Debris-flow Hazards and Related Phenomena, Praxis/Springer, Berlin. pp 489-518

44. Guffanti M, Brantley SR, Cervelli PF, Nye CJ, Serafino GN, Siebert L, Venezky DY, Wald L (2007) Technical-information products for a national volcano early warning system.

45. Hall M (1992) The 1985 Nevado del Ruiz eruption—scientific, social, and governmental response and interaction before the event, Chapter 6. In: McCall G, Laming D, Scott S (eds) Geohazards—Natural and Man-Made, Chapman and Hall, London. pp 43-52

46. Haynes K, Barclay J, Pidgeon N (2007) Volcanic hazard communication using maps: an evaluation of their effectiveness. Bull Volcanol 70:123-138

47. Haynes K, Barclay J, Pidgeon N (2008) The issue of trust and its influence on risk communication during a volcanic crisis. Bull Volcanol 70:605-621

48. Hicks A, Simmons P, Loughin S (2014) An interdisciplinary approach to volcanic risk reduction under conditions of uncertainty. Nat Hazards Earth Syst Sci 14:1871-1887

49. Hoblitt RP, Walder JS, Driedger CL, Scott KM, Pringle PT, Vallance JW (1998) Volcano hazards from Mount Rainier, Washington, Revised 1998.

50. Holtz RD, Schuster RL (1996) Stabilization of slopes. In: Turner AK, Schuster RL (eds) Landslides: Investigation and Mitigation, pp. 439-473

51. Huebl J, Fiebiger G (2005) Debris-flow mitigation measures (Chapter 18). In: Jakob M, Hungr O (eds) Debris-flow Hazards and Related Phenomena, Praxis/Springer, Berlin. pp 445-487

52. Hungr O, Morgan GC, VanDine DF, Lister DR (1987) Debris flow defenses in British Columbia. In: Costa JE, Wieczorek GF (eds) Debris Flows/Avalanches: Process, Recognition, and Mitigation, pp. 201-222

53. (1995) Understanding Volcano Hazards (video).

54. (1996) Reducing Volcanic Risk (video).

55. Janda RJ, Daag AS, Delos Reyes PJ, Newhall CG, Pierson TC, Punongbayan RS, Rodolfo KS, Solidum RU, Umbal JV (1996) Assessment and response to lahar hazard around Mount Pinatubo, 1991 to 1993. In: Newhall CG, Punongbayan RS (eds) Fire and Mud: Eruptions and Lahars of Mount Pinatubo, Philippines, Philippine Institute of Volcanology and Seismology, Quezon City, and University of Washington Press, Seattle. pp 107-139

56. (1988) Sabo works on active volcanoes in Japan. Japan Sabo Assoc, Tokyo.

57. John DA, Sisson TW, Breit GN, Rye RO, Vallance JW (2008) Characteristics, extent and origin of hydrothermal alteration at Mount Rainier Volcano, Cascades Arc, USA: Implications for debris-flow hazards and mineral deposits. J Volcanol Geotherm Res 175:289-314

58. Johnson PA, McCuen RH (1989) Slit dam design for debris flow mitigation. J Hydraul Eng ASCE 115:1293-1296

59. Lagmay AMF, Rodolfo KS, Siringan FP, Uy H, Remotigue C, Zamora P, Lapus M, Rodolfo R, Ong J (2007) Geology and hazard implications of the Maraunot notch in the Pinatubo Caldera, Philippines. Bull Volcanol 69:707-809

60. LaHusen R (2005) Debris-flow instrumentation (Chapter 12). In: Jakob M, Hungr O (eds) Debris-flow hazards and related phenomena, Praxis-Springer, Berlin. pp 291-304

61. Lane LR, Tobin GA, Whiteford LM (2003) Volcanic hazard or economic destitution: hard choices in Baños, Ecuador. Environ Haz 5(1–2):23-34

62. Lavigne F, De Coster B, Juvin N, Flohic F, Gaillard J-C, Texier P, Morin J, Sartohadi J (2008) People's behaviour in the face of volcanic hazards: Perspectives from Javanese communities, Indonesia. J Volcanol Geotherm Res 172(3–4):273-287

63. Leonard GS, Johnston DM, Paton D, Christianson A, Becker J, Keys H (2008) Developing effective warning systems: Ongoing research at Ruapehu volcano, New Zealand. J Volcanol Geotherm Res 172:199-215

64. Liu KF, Chen SC (2003) Integrated debris-flow monitoring system and virtual center. In: Rickenmann F, Chen CL (eds) Debris-Flow Hazards Mitigation: Mechanics, Prediction, and Assessment (Proceedings of the Third International Conference in Davos, Switzerland), Mill press, Rotterdam. pp 767-774

65. Loughlin SC, Baxter PJ, Aspinall WP, Darroux B, Harford CL, Miller AD (2002) Eyewitness accounts of the 25 June 1997 pyroclastic flows and surges at Soufrière Hills Volcano, Montserrat, and implications for disaster mitigation. Geol Soc London, Memoir 21:211-230

66. Major JJ, Janda RJ, Daag AS (1996) Watershed disturbance and lahars on the east side of Mount Pinatubo during the mid-June 1991 eruptions. In: Newhall CG, Punongbayan RS (eds) Fire and Mud: Eruptions and Lahars of Mount Pinatubo, Philippines, Philippine Institute of Volcanology and Seismology, Quezon City, and University of Washington Press, Seattle. pp 895-919

67. Manville VR (2004) Paleohydraulic analysis of the 1953 Tangiwai lahar: New Zealand's worst disaster. Acta Vulcanol 16:137-152

68. Manville VR, Cronin SJ (2007) Breakout lahar from New Zealand's crater lake. EOS Trans Am Geophys Union 88(43):441-442

69. Manville VR, White JDL, Houghton BF, Wilson CJN (1999) Paleohydrology and sedimentology of a post-1.8 ka breakout flood from intracaldera Lake Taupo, North Island, New Zealand. Geol Soc Am Bull 111:1435-1447

70. Marcial S, Melosantos AA, Hadley KC, LaHusen RG, Marso JN (1996) Instrumental lahar monitoring at Mount Pinatubo. In: Newhall CG, Punongbayan RS (eds) Fire and Mud: Eruptions and Lahars of Mount Pinatubo, Philippines, University of Washington Press, Seattle, WA. pp 1015-1022

71. McGuire WJ, Solana MC, Kilburn CRJ, Sanderson D (2009) Improving communication during volcanic crises on small, vulnerable islands. J Volcanol Geotherm Res 183:63-75

72. Mei ETW, Lavigne F, Picquout A, de Bélizal E, Brunstein D, Grancher D, Sartohadi J, Cholik N, Vidal C (2013) Lessons learned from the 2010 evacuations at Merapi volcano. J Volcanol Geotherm Res 261:348-365

73. Michel-Kerjan EO (2010) Catastrophe economics: the National Flood Insurance Program. J Econ Perspect 24:165-186

74. Mileti DS (1999) Disasters by Design: A Reassessment of Natural Hazards in the United States. Joseph Henry Press, Washington, DC.

75. Mileti DS, Sorenson JH (1990) Communication of Emergency Public Warnings: A Social Science Perspective and State-of-the-Art Assessment. Report ORNL-6609. Oak Ridge National Laboratory, Oak Ridge, TN.

76. (1995) Slope Stabilization and Erosion Control—A Bioengineering Approach. Chapman and Hall, London.

77. Mothes PA, Hall ML, Janda RJ (1998) The enormous Chillos Valley lahar: an ash-flow-generated debris flow from Cotopaxi Volcano, Ecuador. Bull Volcanol 59:233-244

78. Myers B, Driedger CL (2008) Eruptions in the Cascade Range during the past 4,000 years: US Geol Surv General Information Product 63 (poster).

79. Myers B, Driedger CL (2008) Geologic Hazards at Volcanoes.

80. Némath K, Cronin SJ (2009) Volcanic structures and oral traditions of volcanism of Western Samoa (SW Pacific) and their implications for hazard education. J Volcanol Geotherm Res 186:223-237

81. Neumann van Padang M (1960) Measures taken by the authorities of the Vulcanological Survey to safeguard the population from the consequences of volcanic outbursts. Bull Volcanol 23(2):181-193

82. Newhall CG, Punongbayan RS (1996) The narrow margin of successful volcanic-risk mitigation. In: Scarpa R, Tilling RI (eds) Monitoring and Mitigation of Volcano Hazards, Springer, Berlin. pp 807-838

83. Newhall C, Aramaki S, Barberi F, Blong R, Calvache M, Cheminee J, Punongbayan R, Siebe C, Simkin T, Sparks S, Tjetjep W (1999) Professional conduct of scientists during volcanic crises. Bull Volcanol 60:323-334

84. (1995) Debris flow at Sakurajima, 2. Kyushu Regional Construction Bureau, Ministry of Construction, Tokyo.

85. O'Shea BE (1954) Ruapehu and the Tangiwai disaster. NZ J Sci Tech B36:174-189

86. Paguican EMR, Lagmay AMF, Rodolfo KS, Rodolfo RS, Tengonciang AMP, Lapus MR, Balisatan EG, Obille EC Jr (2009) Extreme rainfall-induced lahars and dike breaching, 30 November 2006, Mayon Volcano, Philippines. Bull Volcanol 71:845-857

87. Parker DJ, Handmer JW (1998) The role of unofficial flood warning systems. J Contingencies Crisis Manag 6:45-60

88. Paton D, Millar M, Johnston D (2001) Community resilience to volcanic hazard consequences. Nat Hazards 24:157-169

89. Paton D, Smith L, Daly M, Johnston D (2008) Risk perception and volcanic hazard -mitigation: Individual and social perspectives. J Volcanol Geotherm Res 172:179-188

90. Perry RW (1985) Comprehensive Emergency Management: Evacuating Threatened Populations. JAI Press, Greenwich, CT.

91. Perry RW, Greene MR (1983) Citizen Response to Volcanic Eruptions: The Case of Mt. St. Helens. Irvington Publishers, New York, NY.

92. Peterson DW (1988) Volcanic hazards and public response. J Geophys Res 93(B5):4161-4170

93. Peterson DW (1996) Mitigation measures and preparedness plans for volcanic emergencies. In: Scarpa R, Tilling RI (eds) Monitoring and Mitigation of Volcano Hazards, Springer, Berlin. pp 701-718

94. Pielke RA Jr (1999) Nine fallacies of floods. Clim Chang 42:413-438

95. Pielke RA Jr (2007) The Honest Broker—Making Sense of Science in Policy and Politics. Cambridge University Press, Cambridge.

96. (2008) Mount Rainier Volcanic Hazards Plan DEM. Pierce County (Washington) Dept. of Emergency Management (working draft).

97. (2014) Volcanic hazard areas, Chapter 18E.60, Title 18E, Pierce County (Washington) Code, adopted 2004—Ordinance No. 2004–57s, 2014 edition.

98. Pierson TC (1985) Initiation and flow behavior of the 1980 Pine Creek and Muddy River lahars, Mount St. Helens, Washington. Geol Soc Am Bull 96:1056-1069

99. Pierson TC (1989) Hazardous hydrologic consequences of volcanic eruptions and goals for mitigative action -- an overview. In: Starosolszky O, Melder OM (eds) Hydrology of Disasters. Proc. World Meteorological Organization Technical Conference, Geneva, November, 1988, James and James, London. pp 220-236

100. Pierson TC (1998) An empirical method for estimating travel times for wet volcanic mass flows. Bull Volcanol 60:98-109

101. Pierson TC, Scott KM (1985) Downstream dilution of a lahar: Transition from debris flow to hyperconcentrated streamflow. Water Resour Res 21:1511-1524

102. Pierson TC, Janda RJ, Thouret JC, Borrero CA (1990) Perturbation and melting of snow and ice by the 13 November 1985 eruption of Nevado del Ruiz, Colombia, and consequent mobilization, flow, and deposition of lahars. J Volcanol Geotherm Res 41:17-66

103. Pierson TC, Major JJ, Amigo A, Moreno H (2013) Acute sedimentation response to rainfall following the explosive phase of the 2008–2009 eruption of Chaitén volcano, Chile. Bull Volcanol 75:723 doi:10.1007/s00445-013-0723-4

104. Prater CS, Lindell MK (2000) Politics of hazard mitigation. Nat Hazards Rev 1:73-82

105. Rodolfo K (1995) Pinatubo and the Politics of Lahar—Eruption and Aftermath, 1991. University of the Philippines Press, Manila.

106. Ronan K, Johnston D (2005) Promoting Community Resilience in Disasters: The Role for Schools, Youth, and Families. Springer, New York.

107. Ronan K, Paton D, Johnston D, Houghton B (2000) Managing societal uncertainty in volcanic hazards—-a multidisciplinary approach. Disast Prev Mgmt 9:339-348

108. Sager JW, Budai CM (1989) Geology and construction of the Spirit Lake outlet tunnel, Mount St. Helens, Washington. In: Galster RW (ed) Engineering Geology in Washington, Washington State Dept of Natural Resources Bull 78, Olympia, Washington. pp 1229-1234

109. Sager JW, Chambers DR (1986) Design and construction of the Spirit Lake outlet tunnel, Mount St. Helens, Washington. In: Schuster RL (ed) Landslide Dams—Processes, Risk, and Mitigation, pp. 42-58

110. Schelling J, Prado L, Driedger C, Faust L, Lovellford P, Norman D, Schroedel R, Walsh T, Westby L (2014) Mount Rainier is an active volcano--are you ready for an eruption?.

111. Schiechtl HM, Stern R (1996) Ground Bioengineering Techniques for Slope Protection and Erosion Control. Blackwell Scientific, Oxford.

112. Scott KM (1988) Origin, behavior, and sedimentology of prehistoric catastrophic lahars at Mount St. Helens, Washington. In: Clifton HE (ed) Sedimentologic Consequences of Convulsive Geologic Events, pp. 23-36

113. Scott W, Pierson T, Schilling S, Costa J, Gardner C, Vallance J, Major J (1997) Volcano hazards in the Mount Hood region, Oregon.

114. Scott KM, Macías JL, Naranjo JA, Rodríguez S, McGeehin JP (2001) Catastrophic debris flows transformed from landslides in volcanic terrains: Mobility, hazard assessment, and mitigation strategies.

115. Sherburn S, Bryan CJ (1999) The eruption detection system: Mt. Ruapehu, New Zealand. Seismol Res Lett 70:505-511

116. Smart GM (1981) Volcanic debris control, Gunung Kelud, East Java. In: Davies TRH, Pearce AJ (eds) Proceedings of Erosion and Sediment Transport in Pacific Rim Steeplands Symposium, Internat Assoc of Hydrol Sci Pub 132, Christchurch, N.Z. pp 604-623

117. Stein AJ (2001) Mud Mountain Dam. The Online Encyclopedia of Washington State History (HistoryLink.org Essay 3584). [http://www.historylink.org/essays/output.cfm?file_id=3584] http://www.historylink.org/essays/output.cfm?file_id=3584 . Accessed 18 Feb 2014

118. Stone J, Barclay J, Simmons P, Cole PD, Loughlin SC, Ramón P, Mothes P (2014) Risk reduction through community-based monitoring: the *vigías* of Tungurahua, Ecuador. J Appl Volcanol 3:11 (online publication preview)

119. Suryo I, Clarke MCG (1985) The occurrence and mitigation of volcanic hazards in Indonesia and exemplified at the Mount Merapi, Mount Kelut and Mount Galunggung volcanoes. Quart J Eng Geol 18:79-98

120. Tayag J, Punongbayan R (1994) Volcanic disaster mitigation in the Philippines—experience from Mt. Pinatubo Disast 18:1-15

121. Theissen MS (1992) The role of geosynthetics in erosion and sediment control. Geotext Geomembr 11:535-550

122. Tobin GA (1995) The levee love affair—a stormy relationship? Water Resour Bull (JAWRA, J Am Water Resour Assoc) 31(3):359-367

123. Tobin G, Whiteford L (2002) Community resilience and volcano hazard: The eruption of Tungurahua and evacuation of the Faldas in Ecuador. Disasters 26:28-48

124. Umbal JV, Rodolfo KS (1996) The 1991 lahars of southwestern Mount Pinatubo and evolution of the lahar-dammed Mapanuepe Lake. In: Newhall CG, Punongbayan RS (eds) Fire and Mud: Eruptions and Lahars of Mount Pinatubo, Seattle, WA, University of Washington Press, Philippines. pp 951-970

125. (1985) Volcanic emergency management. United Nations, Office of the Disaster Relief Coordinator, New York.

126. Usamah M, Haynes K (2012) An examination of the resettlement program at Mayon Volcano: what can we learn from sustainable volcanic risk reduction? Bull Volcanol 74:839-859

127. (1996) Perilous Beauty—The Hidden Dangers of Mount Rainier (video).

128. (1998) At Risk: Volcano Hazards from Mount Hood, Oregon (video).

129. (2010) Mount St. Helens: A Catalyst for Change. Video program, 6 min 46 sec.

130. (2013) Monitoring Lahars at Mount Rainier.

131. Valentin C, Poesen J, Li Y (2005) Gully erosion: Impacts, factors, and control. Catena 63:132-153

132. Vallance JW (2000) Lahars. In: Sigurdsson H, Houghton BF, McNutt SR, Rymer H, Stix J (eds) Encyclopedia of Volcanoes, Academic Press, San Diego. pp 601-616

133. Vallance JW, Scott KM (1997) The Osceola Mudflow from Mount Rainier: Sedimentology and hazard implications of a huge clay-rich debris flow. Geol Soc Am Bull 109:143-163

134. Voight B (1990) The 1985 Nevado del Ruiz volcano catastrophe: Anatomy and retrospection. J Volcanol Geotherm Res 44:349-386

135. Voight B (1996) The management of volcano emergencies: Nevado de Ruiz. In: Scarpa R, Tilling RI (eds) Monitoring and Mitigation of Volcano Hazards, Springer, Berlin. pp 719-769

136. Watters RJ, Zimbelman DR, Bowman SD, Crowley JK (2000) Rock mass strength assessment and significance to edifice stability, Mount Rainier and Mount Hood, Cascade Range volcanoes. Pure Appl Geophys 157:957-976

137. Waythomas CF, Pierson TC, Major JJ, Scott WE (2013) Voluminous ice-rich lahars generated during the 2009 eruption of Redoubt Volcano, Alaska. J Volcanol Geotherm Res 259:389-413

138. Willingham WF (2005) The Army Corps of Engineers' short-term response to the eruption of Mount St. Helens. Oregon Hist Quart 106:174-203

139. Wisner B, Blaikie P, Cannon T, Davis I (2004) At Risk—Natural Hazards, People's Vulnerability and Disasters. Routledge, New York.

140. Witham C (2005) Volcanic disasters and incidents—a new database. J Volcanol Geotherm Res 148:191-233

141. Wood NJ, Soulard CE (2009) Community exposure to lahar hazards from Mount Rainier, Washington.

142. Wood N, Soulard C (2009) Variations in population exposure and sensitivity to lahar hazards from Mount Rainier, Washington. J Volcanol Geotherm Res 188:367-378

Volcanic Ashfall Preparedness Poster Series: A Collaborative Process for Reducing the VulneraSbility of Critical Infrastructure

Thomas M Wilson[1], Carol Stewart[1,2], Johnny B Wardman[1], Grant Wilson[1], David M Johnston[1,2], Daniel Hill[1], Samuel J Hampton[1], Marlene Villemure[1], Sara McBride[2],Graham Leonard[2,3], Michele Daly[3], Natalia Deligne[3], and Lisa Roberts[4]

[1]Volcanic Ash Testing Lab, Department of Geological Sciences, University of Canterbury, Private Bag 4800, Christchurch 8140, New Zealand

[2]Joint Centre for Disaster Research, Massey University/GNS Science, Wellington, New Zealand

[3]GNS Science, Lower Hutt, New Zealand

[4]Auckland Engineering Lifelines Group Project Coordinator, Infrastructure Decisions Limited, Auckland, New Zealand

ABSTRACT

Volcanic ashfall can be damaging and disruptive to critical infrastructure including electricity generation, transmission and distribution networks, drinking-water and wastewater treatment plants, roads, airports and communications networks. There is growing evidence that a range of preparedness and mitigation strategies can reduce ashfall impacts for critical infrastructure organisations. This paper describes a collaborative process used to create a suite of ten posters designed to improve the resilience of critical infrastructure organisations to volcanic ashfall hazards. Key features of this process were: 1) a partnership between critical infrastructure managers and other relevant government agencies with volcanic impact scientists, including extensive consultation and review phases; and 2) translation of volcanic impact research into practical management tools. Whilst these posters have been developed specifically for use in New Zealand, we propose that this development process has more widely applicable value for strengthening volcanic risk resilience in other settings.

INTRODUCTION

Volcanic ashfall can cause a range of societal impacts. Ashfalls of just a few mm can be damaging and disruptive to critical infrastructure services (also known as 'utilities' in some countries), such as electricity generation, transmission and distribution networks, drinking-water and wastewater treatment plants, roads, airports and communication networks (Wilson et al. [2012b]). Disruption of service delivery can have cascading impacts on wider society. Ashfall can be very widely distributed, potentially affecting communities hundreds of kilometres from the erupting volcano. For example, the recent June 2011 eruption of Puyehue-Cordón Caulle volcanic complex, in southern Chile, deposited ashfall over approximately 75,000 km² of Argentinian Patagonia (Buteler et al. [2011]), with a substantial depth of 30–45 mm deposited on the major regional centre of San Carlos de Bariloche, population approximately 113,000. This led to extensive disruption of the city's water supply, electricity distribution and generation networks, wastewater networks, ground and air transportation networks, and necessitated a major ash clean-up operation within the town (Wilson et al. [2012c]). Specific impacts of ashfall vary considerably, depending on factors such as plant or network design, ashfall characteristics (e.g. loading, grain-size, composition and levels of leachable elements), and environmental conditions before and after the ashfall.

There is also growing evidence that a range of preparedness and mitigation strategies can reduce ashfall impacts (Wardman et al. [2012a], Sword-Daniels et al. [2014]). Core components of disaster risk reduction includes (1) providing advice on likely impacts and best-practice mitigation strategies, and (2) encouraging communities or organisations to adopt preparedness measures which increase their ability to manage hazard consequences, and thus increasing their capacity to manage risk (Paton et al. [2008]; UNISDR [2011]). However, this is not as simple as it seems. Firstly, empowering society to utilise scientific and technological advances to reduce the impacts of disasters is a well-established challenge (Tobin and Montz [1997]; Miletti [1999]; Alexander [2007]; ICSU [2003], [2010]; UNISDR [2011]; Few and Barclay [2011]; McBean [2012]). Both the UNISDR Hyogo Framework for Action (HFA) and Integrated Research on Disaster Risk (IRDR) program to call for more integration of research with the needs of policy and decision makers (ICSU [2008]). Few and Barclay ([2011]) also stress the need to promote integrated, inter-disciplinary approaches, strengthen two-way links between research providers and end-users, and increase experimentation with research mechanisms (such as 'embedded' approaches) to support more effective research/ end-user partnerships.

Secondly, a review of recent risk perception and preparedness studies by Wachinger et al. ([2013]) suggests that even if an individual perceives a high level of risk from a given hazard, this does not necessarily translate into this individual adopting appropriate risk mitigation behaviour for that given hazard. Ballantyne et al. ([2000]) found that provision of hazard information by agencies can, paradoxically, decrease a community's perceived need to prepare as they will tend to transfer responsibility to these agencies. In the case of volcanic hazards, knowledge of proximity to volcanic hazards or susceptibility to their consequences does not assure mitigative actions will be taken, and preparedness levels often remain low in proximal regions even in developed countries (Paton et al. [2008]). Even experiencing a volcanic eruption may not necessarily act as a catalyst for preparing for a future event (Johnston et al. [1999]). These effects may be even more pronounced as eruptions are relatively infrequent and 'exotic' (Paton et al. [1998]). For risk communication, simply providing information often fails to change risk perception or motivate volcanic hazard preparedness, implying that more engaged and appropriate strategies are required (Paton et al. [2008]). This may be overcome by a more participatory process (Twigg [2007]). When stakeholders (e.g. communities and organisations) actively participate as legitimate partners in the communication (and mitigation) exercise, they are empowered to make change which increases their resilience (Covello and Allen [1988]; Paton et al. [2005]). The communication exercise should also be transparent and led by a source which is authoritative, credible and trusted (Berlo [1960]).

This paper describes an example of this approach, a collaborative process used to develop a suite of posters which summarise the potential impacts of volcanic ash and preparedness and mitigation strategies for different sectors of critical infrastructure. The effort included sustained exchange and development of best practices through collaboration among researchers, infrastructure managers and emergency managers from within an established practitioner-research volcanic impact advice structure in New Zealand.

EVOLUTION OF VOLCANIC EMERGENCY MANAGEMENT STRUCTURES IN NEW ZEALAND

Context: 1995–96 Ruapehu Eruption Sequence

Over the past two decades there has been growing awareness in New Zealand (as for many other nations) that volcanic hazards can cause substantial and unique impacts on critical infrastructure (known as 'lifelines' in New Zealand). Consequently, a strong culture of natural hazard risk management within the critical infrastructure sector in New Zealand has developed, catalysed through the development of 'regional engineering lifeline groups' which are defined as "an informal, regionally-based process of lifeline utility representatives working with scientists, engineers and emergency managers to identify interdependencies and vulnerabilities to regional scale emergencies. This collaborative process provides a framework to enable integration of asset management, risk management and emergency management across utilities." (NELC, [2007]). Typically seismic, storm and flood hazards have been the focus, with well-established, evidence-based design codes and advice for preparedness and response strategies available (e.g. the Civil Defence and Emergency Management Act 2002, the Building Act 2004 and the Resource Management Act 1991 of the New Zealand Parliament). By comparison, volcanic hazards have received less attention. This disparity can be at least partially attributed to few, damaging volcanic events occurring during the past 60 years in New Zealand (OCDESC, [2007]). However the 1995–96 eruption of Ruapehu volcano caused widespread and costly impacts to a range of critical infrastructure organisations in New Zealand, despite the relatively modest eruption magnitude (Johnston et al. [2000]). The risk of lahars, blasts and surges closed all three ski fields on Ruapehu for many months while volcanic ashfall and lahars impacted critical infrastructure, agriculture and communities many tens to hundreds of kilometres from the volcano. Total losses were an estimated NZ$130 million (~NZ$188 million or US$161 million in 2014) (Johnston et al. [2000]). Analysis of the performance of responding organisations (national, regional and local government agencies, utilities, emergency services and private businesses) by Paton et al. ([1998]) found there was insufficient knowledge of volcanic hazard impact and appropriate mitigation strategies within these organisations. Nor was there sufficient access to information, which further exacerbated uncertainty regarding preparedness, response and mitigation decision-making. Many organisations looked to the government volcano monitoring agency (formerly the Institute of Geological and Nuclear Sciences, now GNS Science) and the universities for specialist impact and mitigation advice. However, there was limited volcanic impact information easily available, either within New Zealand or internationally (Johnston et al. [2000]).

The Ruapehu crisis was exacerbated by relatively rigid, top-down, siloed management structures at local and regional levels which did not cope well with the impacts occurring across a complex multi-jurisdictional setting (Paton et al. [1998]). In particular, pre-existing networks between information providers and responders were found to be incomplete and inconsistent with respect to information needs.

This detracted from effective communication between organisations hampered decision-making and coordination in an environment characterised by multi-organisational involvement and conflicting demands (Paton et al. [1998]).

Most organizations emerged from the Ruapehu disaster relatively unaffected, and many perceived that they had coped effectively with the demands of the disaster. However Paton et al. ([1998]) argued that this may "stimulate overestimation of future response capability, underestimation of risk, and constrain thinking about future events, making it difficult to conceptualise alternative demands, problems or outcomes...and may ignore the negative outcomes or potential inadequacies of crisis management systems." These authors argued that it was important to ensure that this experience did not result in complacency about future response effectiveness.

Implementing Lessons Learned

In the five to ten years after the Ruapehu eruptions, New Zealand's approach to emergency management has evolved from a 'civil defence' approach to a 'comprehensive emergency management' approach with the passage of the Civil Defence and Emergency Management (CDEM) Act in 2002. This act recognised the unique challenges of managing disasters and emergencies, and stipulated a more coordinated, integrated approach which focused on developing partnerships and clarifying emergency management responsibilities of critical infrastructure companies.

In this changing environment, the lessons from the 1995–96 Ruapehu eruption acted as a catalyst for 1) developing a volcanic impact evidence base to inform preparedness and mitigation decision-making (particularly for ashfall as the most frequently-produced and widely-distributed volcanic hazard); and 2) enhancing communication and coordination structures between volcano and risk scientists and stakeholders (Paton et al. [1998]; Johnston et al. [2000]; Leonard et al. [2008]).

Volcanic Impacts Research Group

As part of New Zealand's increased investment in applied volcanology research over the past 15 years, a volcanic impact research group was formed between GNS Science and partner universities (University of Canterbury, Massey University, and University of Auckland). This group has pursued a sustained and systematic approach to assessing the impact of volcanic ash on critical infrastructure, for as wide a range of different eruption types and magnitudes. This group has undertaken reconnaissance trips to areas impacted by volcanic eruptions worldwide at varying intervals after the eruption, to capture both short and longer term impacts, timescales of recovery, successful mitigation strategies and overall management lessons (Table 1). A further goal is to develop quantitative risk tools, such as vulnerability and fragility functions that relate impacts to the amount and characteristics of ashfall received and to develop more quantitative relationships for use in risk modelling. The group has also studied cascading impacts of ashfall within a systems-thinking framework (Wilson et al. [2012b]; Sword-Daniels et al. [2014]). More recently, empirical laboratory-based testing of critical infrastructure components has been conducted in our Volcanic Ash Testing Lab (VAT Lab) (Wilson et al. [2012a]; Wardman et al. [2012b]). The strategic focus of the full research group has been on understanding both ashfall impacts on individual system components and overall system functionality. The group has received ongoing funding support from the Natural Hazard Resource Platform (a multi-party research platform funded by the New Zealand Government dedicated to increasing New Zealand's resilience to Natural Hazards via high quality collaborative research), critical infrastructure organisations (primarily AELG organisations, described in the following section), and the New Zealand Earthquake Commission. In kind funding support from.

Table 1: List of volcanic impact reconnaissance trips undertaken by New Zealand volcanic impact research group

Volcano	Country	Year of eruption	Year of assessment trip
Mt St Helens	USA	1980	1995
Crater Peak (Mt Spurr)	USA	1989	1996
Sakura-jima	Japan	~1980-2000	2001

Ruapehu	New Zealand	1995-96	1995-97
Etna	Italy	2003	2003 (several days after)
Tungurahua & Reventador	Ecuador	1999-2005 & 2002	2005
Merapi	Indonesia	2006	2006 (1 month after)
Pinatubo	Philippines	1991	2007
Eldfell	Iceland	1973	2008
Hudson	Chile	1991	2008
Chaiten	Chile	2008	2009
Redoubt	USA	2009	2009
Pacaya	Guatemala	2010	2010 (4 months after)
Tungurahua	Ecuador	2010	2010 (4 months after)
Shinmoedake	Japan	2011	2011 (9 months after)
Puyehue-Cordón Caulle	Chile	2011	2012 (9 months after)
Tongariro	New Zealand	2012	2012 (2–3 days after)

Wilson et al.

Wilson et al. Journal of Applied Volcanology 2014 3:10, doi: 10.1186/s13617-014-0010-x

Provision and Coordination of Volcanic Impact Knowledge

In conjunction with development of the research group, an enhanced partnership with end-users needed to be established for communicating volcanic impact science both during crisis and non-crisis periods. The Auckland Engineering Lifelines Group (AELG) is a group of critical infrastructure organisations within the Auckland region. Its mission is to increase critical infrastructure resilience to all hazards. As such, there was considerable interest within AELG to enhance volcanic impact science capability. Volcanic hazards are one of the most substantial risks to the Auckland region, either from an eruption from the Auckland Volcanic Field upon which the city is built or from distal ashfall hazards from central North Island volcanoes. In 2004, the Volcanic Impact Study Group (VISG) was established as a sub-committee of the AELG. The VISG was designed to be a multidisciplinary and multi-institution consortium of volcanology and natural hazard researchers and practitioners with the following aims (VISG [2012]):

To collate and advocate existing knowledge about the impacts of volcanic hazards (e.g. ash) on, and mitigation measures for, lifeline infrastructure.

To facilitate and support research on the impacts of volcanic hazards on lifelines and people, and the development of appropriate mitigation measures.

To provide input into the applicability for lifelines of any research being undertaken.

To facilitate reconnaissance investigations, and/or advocate lifeline representation on reconnaissance investigations, to active volcanic areas where this would add to our knowledge about volcanic impacts on infrastructure.

To provide a national focal point for volcanic impacts work on lifelines.

Initially, the VISG was only focused on the Auckland region and was concerned primarily with impacts from the Auckland volcanic field, an active basaltic scoria cone field upon which Auckland City (pop. 1.5 million) is constructed (Lindsay et al. [2010]). This focus has since broadened to support volcanic impacts research with any local, regional or national stakeholder in New Zealand. VISG provides a formalised networked structure between volcanic impact science providers (GNS and the universities) and critical infrastructure and emergency management organisations. Key activities of the VISG include undertaking focused research on volcanic impacts, contributing to volcano contingency planning and exercising when requested, and running an annual seminar on current research.

Communication of appropriate volcanic impact science with end-users in a timely manner during an eruption crisis can be additionally challenging in the absence of adequate training and communication structures linked to expert knowledge. Pre-existing relationships between end-users and researchers, combined with readily available resources, can greatly reduce information searching and processing time, which aids decision-making timeliness and quality (Paton et al. [1998]). The VISG aims to improve non-crisis and crisis communication between providers and recipients by developing relationships and resources which anticipate and provide for likely information needs. It fosters a group of information providers who can access, collate, interpret and disseminate information as needed within a known and regularly used framework. Likewise, the interaction with AELG and other lifeline group members contributes to developing a capacity within their own organisations to interpret, request and use specialist volcanic impact information.

Specific activities have included multi-organisation workshops, targeted 'sector specific' workshops, one-on-one meetings and public lecture tours. Information is also provided to international volcanological initiatives, such as the USGS-GNS Volcanic Ash Impacts Website (http://volcanoes.usgs.gov/ash/), the International Volcanic Health Hazard Network (www.ivhhn.org) and the Cities on Volcanoes Commission of IAVCEI (http://cav.volcano.info/).

POSTER DESIGN

Critical infrastructure organisations that have experienced adverse impacts during ashfall events commonly report low levels of prior awareness of ashfall hazards and impacts (Blong [1984]; Paton et al. [1998]; Ronan et al. [2000]; Wilson et al. [2012b]). Whilst many organisations recognise the value of planning and preparedness for volcanic hazards, the necessary investment can be difficult to justify in the context of a variety of other hazards and business pressures. Feedback from AELG members suggested that lengthy reports summarising known impacts, mitigation options/recommendations and interdependency issues were useful, but only during infrequent detailed planning exercises. Authoritative but concise reference materials preferred as a means to inform planning and be readily available during a crisis, supplemented by additional information from science providers as needed. After some experimentation and consultation, posters were judged to be the optimal method for condensing key impact and mitigation information into a concise, palatable and visible form. The first series of posters was commissioned and completed during the period 2007–2010 for five infrastructure sectors: airports, road networks, drinking-water supplies, power-systems (networks), and wastewater collection and treatment systems (Figures 1, 2, 3, 4, 5). These sectors were selected by AELG and VISG members as most likely to be impacted. This edition of the posters were advertised widely in outreach activities, used during emergency management exercises and ultimately became a recognised information source in New Zealand (Bay of Plenty Engineering Lifelines Coordinator pers. comm. 2012).

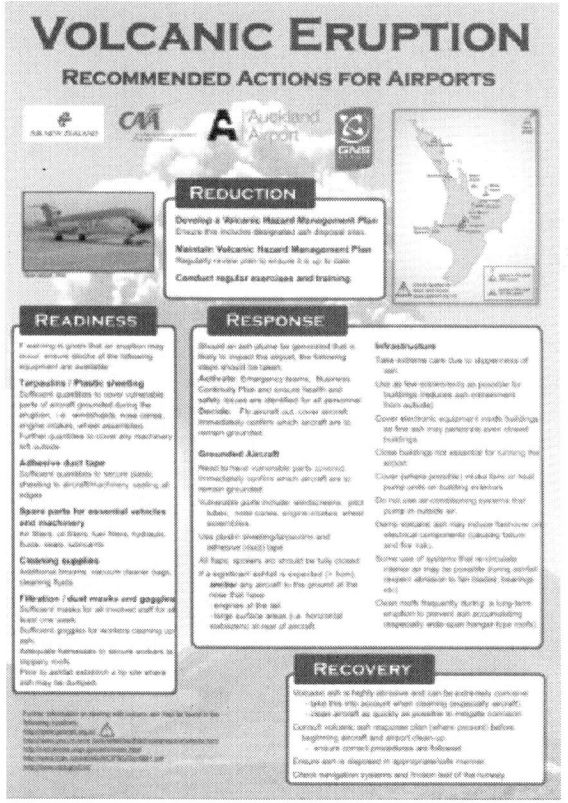

Figure 1: Recommended actions for airports to mitigate ashfall hazard.

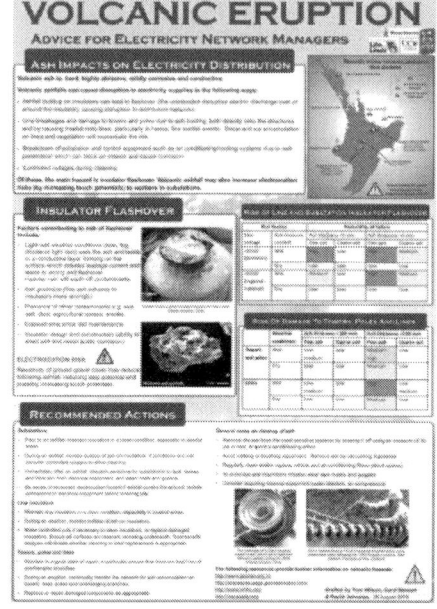

Figure 2: Recommended actions for Electricity Network Managers to mitigate ashfall hazard.

Figure 3: Recommended actions for Road Managers to mitigate ashfall hazard.

Figure 4: Recommended actions for Water Supply Managers to mitigate ashfall hazard.

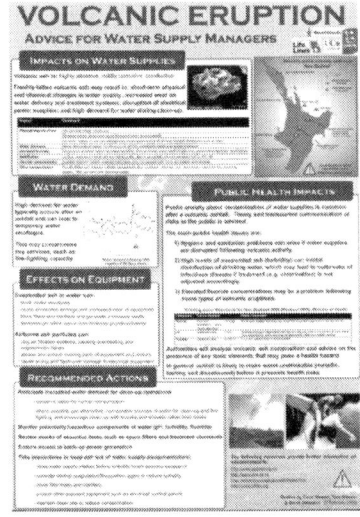

Figure 5: Recommended actions for Waste Water Managers to mitigate ashfall hazard.

During subsequent review of VISG resources and risk communication strategy, it became apparent that the content of the first series of posters was becoming outdated; for example, global initiatives in the aviation sector (ICAO, [2007]) needed to be incorporated into advice. Thus it was decided in 2012 that a) the current poster suite should be updated with the latest research and accounting for local and global developments, and b) that additional posters should be developed to address knowledge gaps. Subjects of particular interest were advice on ash cleanup operations for city authorities; impacts on building facilities; impacts on heating, ventilation and air-conditioning (HVAC) systems and emergency power generators; and impacts on computers and electronics. A further change was that the content of the original poster on power systems was split between two new posters: one on electricity generation facilities and the other on electricity transmission and distribution networks. This expansion made it possible to incorporate substantial new research in this area (Wardman et al. [2012a]). The new series of posters are shown in Figures 6, 7, 8, 9, 10, 11, 12, 13, 14, and 15. We note that despite telecommunications being a key critical infrastructure sector, we did not consider there to be sufficient documentation of impacts or mitigation guidance to create a robust poster.

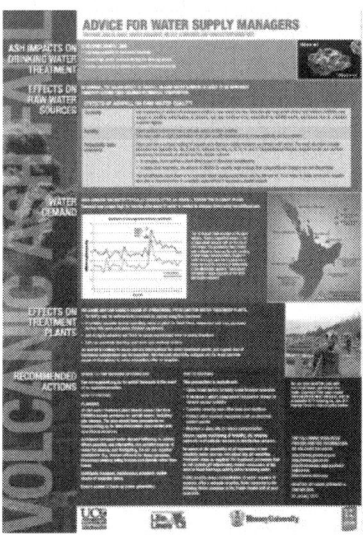

Figure 6: Volcanic Ash: Advice for Water Supply Managers.

Figure 7: Volcanic Ash: Advice for Wastewater Managers.

Figure 8: Volcanic Ash: Advice for Power Plant Operators.

Figure 9: Volcanic Ash: Advice for Power Transmission and Distribution System Operators.

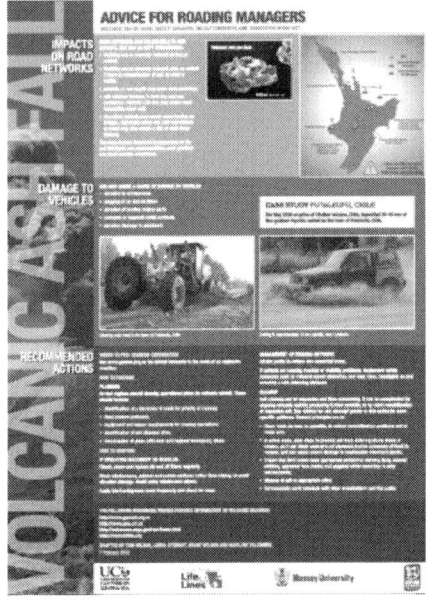

Figure 10: Volcanic Ash: Advice for Roading Managers.

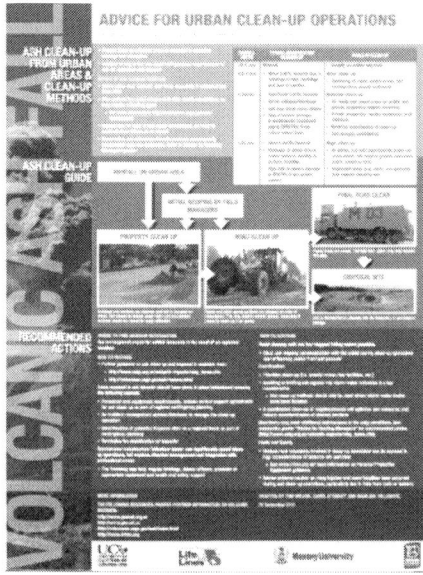

Figure 11: Volcanic Ash: Advice for Urban Clean-up Operations.

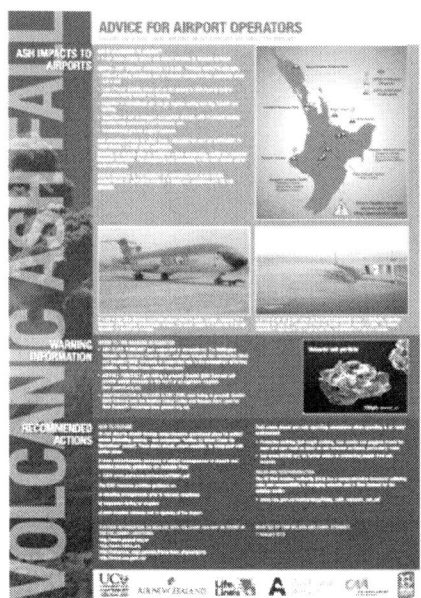

Figure 12: Volcanic Ash: Advice for Airport Operators.

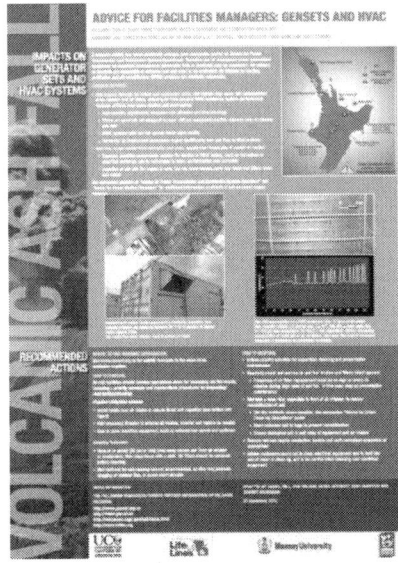

Figure 13: Volcanic Ash: Advice for Facilities Managers: GenSets and HVAC.

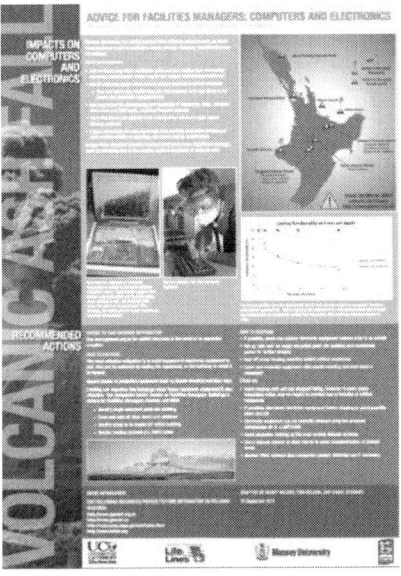

Figure 14: Volcanic Ash: Advice for Facilities Managers: Computers and Electronics.

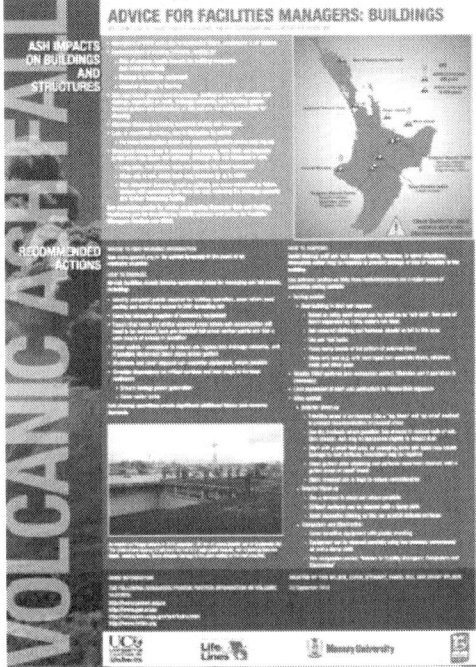

Figure 15: Volcanic Ash: Advice for Facilities Managers: Buildings.

Poster Content

Content was derived from the research team's observations of the consequences of volcanic eruptions around the world (summarised in Wilson et al. [2012b]). These insights were supplemented by findings from empirical laboratory experiments, such as the vulnerability of high-voltage transmission insulators to flashover from volcanic ash contamination (e.g. Wardman et al. [2012a]; [2012b]; Wilson et al. [2012b]). Poster content was written to be practical, with action-based knowledge. Expert elicitation from AELG members was used to ensure that content was technically correct, relevant and used accurate with terminology (Figure 16). Consultation broadened beyond AELG as required: power generating companies within Bay of Plenty Engineering Lifelines Group contributed to and reviewed 'Advice for Power Plant Operators', the Ministry of Health reviewed 'Advice for Water Supply Managers' and the Civil Aviation Authority reviewed 'Advice for Airport

Managers'. Active involvement with the Ministry of Health has also contributed to improved volcanic health impact coordination between volcanic impact scientists and public health professionals. This approach ensured access to the best possible knowledge, facilitated broad participation of relevant organisations, increased awareness of the posters as a resource, and raised the visibility of VISG.

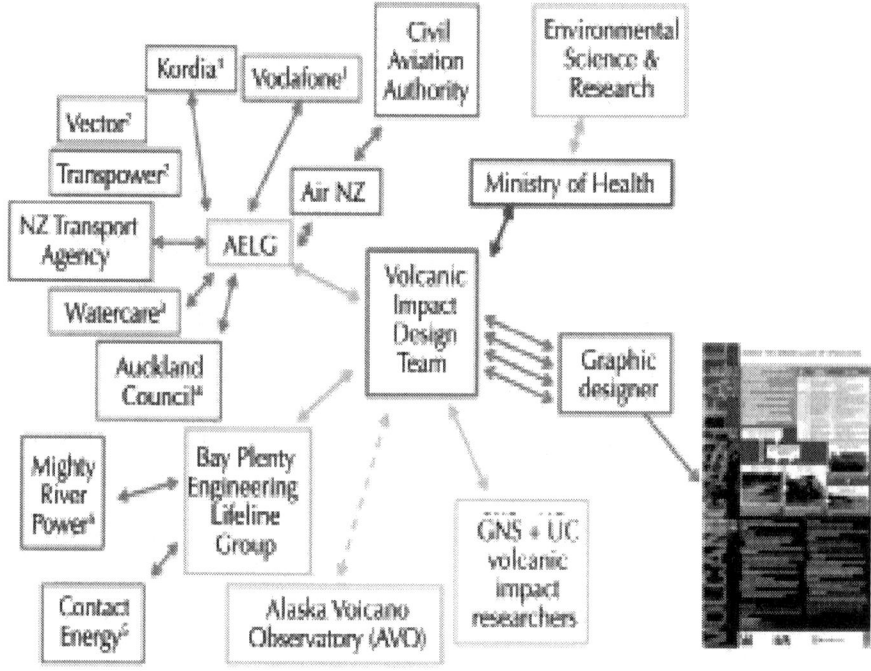

Figure 16: Diagram summarising extent of consultation process for volcanic impact and mitigation posters. Blue: critical infrastructure organisations who chose to actively participate in process; Green: critical infrastructure advisory groups; Red: poster design team; Purple: government agency; Yellow: science groups (both government and academic).

Posters are tailored for individual sectors and reflect each sector's approach to volcanic risk management. Therefore the 'Advice for Airport Managers' poster simply summarises likely impacts and directs airport managers towards national and global planning and response resources, such as the International Civil Aviation Organisation (ICAO) reference guides. The involvement of Air New Zealand Ltd (the major regional airline in New Zealand) and the New Zealand Civil Aviation Authority in designing and reviewing the poster was essential to create a resource aligned with industry standards and suitable for the New Zealand aviation sector.

The restricted space in a poster format enforced concise summaries of impacts and mitigation measures. It was therefore important to be able to refer to further resources and the posters were designed to link with established, authoritative volcanic ash impact information sources. The USGS/GNS volcanic ash impacts website (http://volcanoes.usgs.gov/ash/) and the International Volcanic Health Hazard Network website (www.ivhhn.org) are referred to on nearly all posters, depending on topic and intended audience. Sector-specific resources are provided where available, such as the ICAO Manual on Volcanic Ash, Radioactive Material and Toxic Chemical Clouds (ICAO [2007]) referenced on the poster on 'Advice for Airport Managers'.

Design

The posters are designed as fact sheets which refer the specialist audience to specific information, such as further web-based resources or industry standards where appropriate. Language, terminology and graphics used on the posters are designed primarily for the target audience of New Zealand critical infrastructure managers. Design elements of the posters are described in Figure 17.

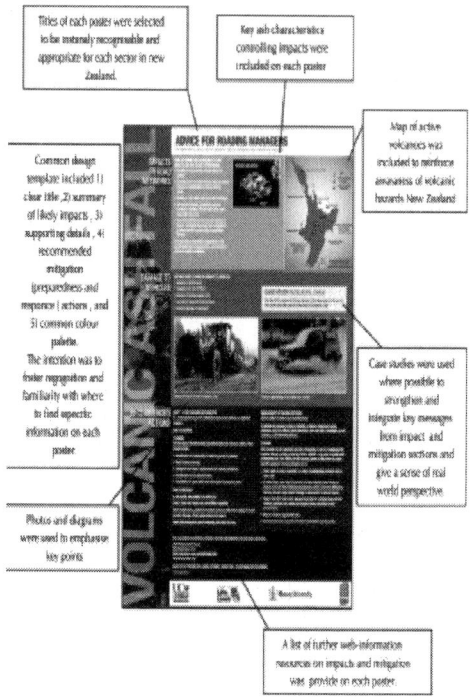

Figure 17: Overview of the current (Series 2; Figures6, 7, 8, 9, 10,11, 12, 13, 14, 15) volcanic impact poster design elements.Compare with Series 1 to see changes (Figures 1, 2, 3, 4, 5).

Review Process

The posters underwent a two-stage review process. Initially they were reviewed by a team of eight scientists within the VISG project team, then submitted to a technical sub-group of the AELG or other appropriate organisations (Figure 16), typically including engineers, risk managers and business continuity advisors. Their feedback was used to revise the posters. This process was repeated as required, with up to five iterations in some cases. Posters were also reviewed by colleagues from the Alaska Volcano Observatory, who have extensive operational experience in responding to ash-producing eruptions and interacting with affected sectors before, during and after ashfall events. This provided a valuable external perspective.

Dissemination

Once finalised, the updated Series 2 posters were distributed to all AELG members, to the New Zealand National Engineering Lifelines Committee for national distribution, and also hosted on the AELG and GNS Science websites as an open access resource (http://www. aelg.org.nz/volcanic-impacts/visg-projects/; http://www.gns.cri.nz/ Home/Learning/Science-Topics/Volcanoes/Eruption-What-to do/Ash-Impact-Posters). Public outreach talks and briefings by GNS Science staff in New Zealand, which regularly include briefings to regional engineering lifeline groups, routinely promote awareness of the posters, along with other preparedness and mitigation resources. An annual volcanic hazard short-course for infrastructure and emergency managers also uses the posters during exercises. They are also used in university teaching for scenario-based role-play simulations. Series 1 posters were also widely disseminated and utilised during the 2012 Te Maari eruption from Tongariro volcano.

The suite of posters has also been shared internationally, via distribution by the IAVCEI Cities and Volcanoes Commission's Volcanic Ash Impacts Working Group and will be hosted on the USGS Volcanic Ash Impacts Website (http://volcanoes.usgs.gov/ash/index.html) as a resource for the global community.

Posters in Action – Esquel Case Study

A practical test of the posters' utility occurred during the May 2008 eruption of Volcan Chaitén, Chile (Stewart et al. [2009]). In early May 2008 widespread ashfall from the explosive rhyolitic eruption was distributed by the prevailing westerly winds over Argentina. The city of Esquel (pop. 35,000), located 110 km east of the volcano in Chubut province, Northern Patagonia, received approximately 5 mm of fine ash on the morning of 5 May (Figure 18A). Public authorities were immediately concerned about contamination of the city's water supply as residents reported a 'strong metallic taste' in the drinking water.

Figure 18: Chaiten ashfall in Esquel, Argentina. A) Approximately 5 mm of fine-grained rhyolitic ash fell in the town on 5 May 2008; B) The Canal de Faldeo open water supply line for Esquel, Argentina.

The water sources for the city are primarily groundwater and thus are relatively resilient to ashfall contamination. However, there is a point of vulnerability where the water is delivered to the treatment plant along the open, concrete-lined 2.3 km-long Canal de Faldeo (Figure 18B).

The water supply authority did not have any knowledge of potential impacts of an ashfall on the water supply. In their search for information they contacted a member of our research team (CS) who had authored a review of the subject (Stewart et al. [2006]). She provided advice, in collaboration with a local university, on an appropriate water sampling and monitoring regime and interpretation of ashfall leachate data. Using the poster "Advice for water supply managers" (Figures 1, 2, 3, 4, and 5), she also provided guidance on impacts and mitigation strategies. Water sampling showed that levels of sulphate and dissolved iron and aluminium were higher in the Canal de Faldeo than the raw water source, and to a lesser extent, in treated drinking water (Stewart et al. [2009]). These elevated levels were sufficient to produce a noticeable taste in the final drinking water but remained well below Argentinian drinking water standards (see Stewart et al. [2009]). The water authority was thus able to reassure the public that ashfall contamination of the water source did not pose a public health risk.

Two-way exchange of information between the poster design team and the water authority was critical for ground-truthing and refining the management advice on the posters. Our predictions were that the primary impacts of the ashfall would be an increase in raw water turbidity and that water demand would increase as residents cleared ash from their properties. These both proved to be the case. Local authorities also noted the value of the poster's advice to communicate information to the public in a timely and transparent manner as the metallic taste in the water had caused some anxiety about contamination of the water supply.

Internationalising Posters?

The case study above illustrates that these posters may be useful tools during an eruption crisis beyond the New Zealand context for which they were designed. The technical and engineering content of the posters was based on findings of ashfall impact assessment trips, to an extensive range of volcanically-active countries (Table 1, volcanic impacts research group). Thus, the advice given is applicable to infrastructure not just in New Zealand (which has highly-modernised infrastructure) but in other, less-developed, settings. For example, the 'Advice for Wastewater Managers' poster (Figure 7) describes ashfall impacts on individual system components, so that individual treatment facilities can select relevant components. Similarly, many components of infrastructure systems such as pumping equipment, HVAC units and engine components are universal thus the mitigation advice given is applicable.

However, we note that the emergency management content of the posters is specific to New Zealand. This includes aspects such as where to find warning information in the event of an eruption, and (for the 'Advice for Airport Managers' poster) contact details for the local Volcanic Ash Advisory Centre (VAAC).

REFERENCES

1. This paper describes a collaborative process used to create a suite of ten informational posters intended to improve the resilience of critical infrastructure organisations to volcanic ashfall hazards.

Key features of this process were: a collaborative partnership between critical infrastructure managers and relevant government agencies with volcanic impact scientists; consultation and review phases; and translation of volcanic impact research into practical management tools.

2. In addition to producing the posters, which are a unique global resource, the process has further enhanced and grown networks between volcanic impact scientists/agencies and critical infrastructure organisations. We note that our work has been developed in a New Zealand context and thus has relied heavily on the highly networked VISG and AELG structures, and existing risk management culture. Whilst the posters have utility beyond New Zealand, as demonstrated by the Chaitén case study, we propose that this development process may be a useful model for strengthening volcanic risk resilience in other settings.

3. AUTHORS' CONTRIBUTIONS

4. TW and CS planned and conducted the research, and wrote the manuscript. JW, GW, DJ, DH, SH, MV and LR contributed to poster content and design. SM, GL, MD, ND sand LR contributed to poster design and review and to manuscript preparation. All authors have read, reviewed and approved the final manuscript.

5. ACKNOWLEDGEMENTS

6. We kindly thank all poster reviewers from the AELG, BoP ELG, NZ Ministry of Health and ESR. Special thanks to Brian Park (WaterCare), Scott Muspratt (Vector Ltd), Bob Fletcher (Air NZ) and Peter Lechner (NZ CAA). Our sincere thanks to Tina Neal and Kristi Wallace from the Alaska Volcano Observatory for their review of posters and continuing support. Thanks also to Jim Cole for supporting the project. We acknowledge funding support from AELG, the DEVORA project (funded by the Earthquake Commission, Auckland Council and Ministry of Business, Innovation and Employment), and MBIE Research Contract C05X0907 (TW, CS, DJ).

7. REFERENCES

8. Alexander D (2007) making research on geological hazards relevant to stakeholders' needs. Quaternary Int 171–172:186-192

9. Ballantyne M, Paton D, Johnston D, Kozuch M, Daly M (2000) Information on volcanic and earthquake hazards: the impact on awareness and preparation. Institute of Geological & Nuclear Sciences Limited Science Report 2000/2, Lower Hutt.

10. Berlo DK (1960) the process of communication. Holt, Rinehart & Winston, New York.

11. Blong RJ (1984) Volcanic Hazards: A Sourcebook on the Effects of Eruptions. Academic Press, Sydney.

12. Buteler M, Stadler T, Lopez Garcia GP, Lassa MS, Trombotto Liaudat D, D'Adamo P, Fernandez-Arhex V (2011) Propiedades insecticidas de la ceniza del complejo volcanic Puyehue-Cordon Caulleny su possible impacto ambiental. Rev Soc Entomol Argent 70:149-156 Abstract in English

13. Covello VT, Allen F (1988) Seven Cardinal Rules of Risk Communication. U.S. Environmental Protection Agency. Policy Document OPA-87-020, Washington, D.C.

14. Few R, Barclay J (2011) Societal Impacts of Natural Hazards, a review of international research funding. UK Collaborative on Development Sciences, London. [www.paris.icao.int/news/pdf/9691.pdf]

15. ICAO (2007) Manual on Volcanic Ash, Radioactive Material and Toxic Chemical Clouds. Doc 9691 AN/954, International Civil Aviation Organization, p 159p. Available from

16. (2003) Priority Area Assessment on Environment and its Relation to Sustainable Development. International Council for Science, Paris.

17. (2008) a Science Plan for Integrated Research on Disaster Risk (IRDR. International Council for Science, Paris.

18. (2010) Regional Environmental Change: Human Action and Adaptation. What does it take to meet the Belmont Challenge? Preliminary report of an ad hoc ICSU panel. International Council for Science, Paris.

19. Johnston DM, Bebbington M, Lai C-D, Houghton BF, Paton D (1999) volcanic hazard perceptions: comparative shifts in knowledge and risk. Disast Prev Manag 8:118-12

20. Johnston DM, Houghton B, Neall VE, Ronan KR, Paton D (2000) Impacts of the 1945 and 1995–1996 Ruapehu eruptions, New

Zealand: an example of increasing societal vulnerability. Geol Soc Am Bull 112(5):720-726

21. Leonard GS, Johnston DM, Paton D, Christianson A, Becker JS, Keys H (2008) Developing effective warning systems: ongoing research at Ruapehu volcano, New Zealand. J Volcanol Geotherm Res 172(3/4):199-215

22. Lindsay J, Marzocchi W, Jolly G, Constantinescu R, Selva J, Sandri L (2010) Towards real-time eruption forecasting in the Auckland Volcanic Field: application of BET_EF during the New Zealand National Disaster Exercise 'Ruaumoko'. Bull Volcanol 72(2):185-204

23. McBean GA (2012) Integrating disaster risk reduction towards sustainable development. Curr Opin Environ Sustain 4:122-127

24. Miletti D (1999) Disasters by Design. Joseph Henry Press, Washington, DC.

25. (2007) National Engineering Lifeline Committee Brochure.

26. (2007) National Hazardscape Report. Officials' Committee for Domestic and External Security Coordination. Department of the Prime Minister and Cabinet, New Zealand.

27. Paton D, Johnston DM, Houghton BF (1998) Organisational response to a volcanic eruption. Disast Prev Manag 7:5-13

28. Paton D, Smith L, Johnston DM (2005) when good intentions turn bad: promoting natural hazard preparedness. Australian J Emerg Manag 20(1):25-30

29. Paton D, Smith L, Daly M, Johnston DM (2008) Risk perception and volcanic hazard mitigation: Individual and social perspectives. J Volcanol Geotherm Res 172:179-188

30. Ronan KR, Paton D, Johnston DM, Houghton BF (2000) Managing societal uncertainty in volcanic hazards: a multidisciplinary approach. Disast Prev Manag 9(5):339-348

31. Stewart C, Johnston DM, Leonard GS, Horwell CJ, Thordarsson T, Cronin SJ (2006) Contamination of water supplies by volcanic ashfall: a literature review and simple impact modelling. J Volcanol Geotherm Res 158:296-306 Publisher Full Text

32. Stewart C, Pizzolon L, Wilson TM, Leonard GS, Dewar D, Johnston DM, Cronin SJ (2009) Can volcanic ash poison water supplies? Integr Environ Assess Manag 5(4):713-716

33. Sword-Daniels V, Wilson TM, Sargeant S, Rossetto T, Twigg J, Johnston DM, Loughlin SC, Cole PD (2014) Consequences of long-term volcanic activity for essential services in Montserrat: challenges, adaptations and resilience. Memoir of the Geological Society of London Special volume "The Eruption of Soufriere Hills Volcano".in press

34. Tobin GA, Montz BE (1997) Natural Hazards. Guilford Press, New York. [https://practicalaction.org/docs/ia1/community-characteristics-en-lowres.pdf]

35. Twigg J (2007) Characteristics of a disaster-resilient community: a guidance note. DFID Disaster Risk Reduction Interagency Coordination Group. Accessed 30 January 2014; Available from:

36. UNISDR (2011) HFA Progress in Asia-Pacific: Regional Synthesis Report 2009–2011. United Nations Office for Disaster Risk Reduction - Regional Office for Asia and Pacific (UNISDR AP), p 37 p [http://www.aelg.org.nz/volcanic-impacts/about-visg/]

37. VISG (2012) Volcanic Impacts Study Group Charter. Auckland Engineering Lifelines Group. Available from; Accessed on 10 August 2013

38. Wachinger G, Renn O, Begg C, Kuhlicke C (2013) the Risk Perception Paradox—Implications for Governance and Communication of Natural Hazards. Risk Anal 33(6):1049-1065

39. Wardman JB, Wilson TM, Bodger PS, Cole JW, Stewart C (2012) Potential impacts from tephra fall to electric power systems: A review and mitigation strategies. Bull Volcanol 74(10):2221-2241

40. Wardman JB, Wilson TM, Bodger PS, Cole JW, Johnston DM (2012) Investigating the electrical conductivity of volcanic ash and its effect on HV power systems. Physics and Chemistry of the Earth, Parts A/B/C 45–46:128-145

41. Wilson G, Wilson TM, Cole JW, Oze C (2012a) Vulnerability of laptop computers to volcanic ash and gas. Nat Hazards 63(2):711-736

42. Wilson TM, Stewart C, Sword-Daniels V, Leonard GS, Johnston DM, Cole JW, Wardman J, Wilson G, Barnard ST (2012b) Volcanic ash impacts on critical infrastructure. Physics and Chemistry of the Earth, Parts A/B/C 45–46:5-23

43. Wilson T, Stewart C, Bickerton H, Baxter P, Outes V, Villarosa G, Rovere E (2012c) The health and environmental impacts of the June 2011 Puyehue-Cordón Caulle volcanic eruption: a preliminary report. GNS Science Report 2012/20, p 27.

Citations

CHAPTER 1

Raquel Otoni de Araújo, Teresa C. da Silva Rosa, Socio-Environmental Vulnerability and Disaster Risk Reduction: The Role of Espírito Santo State (Brazil), http://dx.doi.org/10.1590/1809-4422ASOC1068V1742014.

CHAPTER 2

Claudia Bach, Sara Bouchon, Alexander Fekete, Jörn Birkmann, and Damien Serre, « Adding value to critical infrastructure research and disaster risk management: the resilience concept », *S.A.P.I.EN.S* [Online], 6.1 | 2013, Online since 15 July 2014, connection on 13 March 2015. URL : http://sapiens.revues.org/1626

CHAPTER 3

Herlander Mata-Lima, Andreilcy Alvino-Borba, Adilson Pinheiro, Abel Mata-Lima, and José António Almeida, Impacts of Natural Disasters on Environmental and Socio-Economic Systems: What Makes the Difference?, http://dx.doi.org/10.1590/S1414-753X2013000300004.

CHAPTER 4

Airton Bodstein; Valéria Vanda Azevedo de Lima; and Angela Maria Abreu de Barros, The Vulnerability of the Elderly in Disasters: the Need for an Effective Resilience Policy, http://dx.doi.org/10.1590/S1414-753X2014000200011.

CHAPTER 5

Gangalal Tuladhar, Ryuichi Yatabe, Ranjan Kumar Dahal, and Netra Prakash Bhandary, Disaster Risk Reduction Knowledge of Local People in Nepal, doi:10.1186/s40677-014-0011-4.

CHAPTER 6

Anna Hicks and Roger Few, Trajectories of social vulnerability during the Soufrière Hills volcanic crisis, doi:10.1186/s13617-015-0029-7.

CHAPTER 7

Angela K Diefenbach, Nathan J Wood, and John W Ewert, Variations in Community Exposure to Lahar Hazards from Multiple Volcanoes in Washington State (USA), doi:10.1186/s13617-015-0024-z.